W9-CJO-068

VETS FOR VETS:

Harnessing the Power of Vets to Heal

VETS FOR VETS:

Harnessing the Power of Vets to Heal

Gerald D. Alpern, Ph.D.

Published in the United States by
Psychological Development Publications.
www.vetsforvets.info

ISBNs:

978-1-62217-830-8 (paperback)

978-1-62217-927-5 (hardcover)

Manufactured in the United States of America

First Edition

Foreword text used with permission from Paul Anderson,
Founder and Executive Director of Huts for Vets.

Testimonials reprinted with permission from Huts for
Vets participants.

Contents

Foreword

Dr. Alpern is one of the most amazing veterans I know. He has dedicated himself to helping veterans thrive. His work over the last six decades with veterans began when he was in the Army during the Korean War.

Vets for Vets represents the culmination of his experience and wisdom, describing appropriate ways to treat our ailing veteran population, while warning against practices that have proven detrimental to our service members. He does this through a comprehensible and easy-to-read approach that is well suited to both laymen and professionals.

This book offers new insights and new ways of helping veterans by addressing three groups of readers:

- Veterans who have all but given up hope of being able to return to productive civilian lives

- Families and friends of veterans who may be at a loss as to how to relate to or help their loved, returning veterans

- Mental health workers who have been frustrated by their lack of success in helping this huge and needy population

A major contribution of this book is Dr. Alpern's message of how individual veterans and groups of veterans can provide unique and effective help for their fellow veterans. He provides specific new treatment techniques for professional mental health workers and agencies.

Dr. Alpern has taken a daring approach by enumerating the many inappropriate and devastating "treatment" techniques prescribed by government agencies. Dr. Alpern then offers successful techniques in which families, friends, communities, and mental health professionals can successfully help suffering veterans overcome both institutional and personal obstacles on their divergent paths toward healthy civilian lives.

I know well of Dr. Alpern's passion for helping veterans through his volunteering with Huts for Vets wilderness programs. He has volunteered serving as the team psychologist on every HFV trip over the last three years. One of our trails crests a mountaintop at nearly 12,000 feet, and Dr. Alpern often outhikes men fifty years younger than he. Dr. Alpern models a life well lived, a mental health professional who is revered as a sage and witty oracle dispensing to post-9/11 veterans his wisdom, guidance, and, above all, his deep caring.

Dr. Alpern in 2014 applied his passion to the theater, where he produced a play, *Make Sure It's Me*, by Kate Wenner. Focused on the common and often untreated and unrecognized veteran ailment of traumatic brain injury (TBI), Dr. Alpern utilized Huts for Vets wilderness trip alumni to act out roles for which they were, sadly and ironically, perfectly cast. The play was a huge hit and and was a healing experience for our veteran actors. Most importantly, audiences discovered a deep sympathy for veterans suffering a plight of which the American public has little knowledge.

Vets For Vets is yet another outpouring of Dr. Alpern's experience and expertise. This book creates a format in which veterans will be empowered to help their peers. With service comes healing, self-esteem, and the establishment of a veterans' network based on mutual aid and caring for those who have given their all in service to their country, their communities, and their families.

Huts for Vets and *Vets For Vets* combine to empower veterans to serve their peers, expand their educational opportunities, and help them find peace and perspective. As a partner in this mission, I am honored to work with Dr. Alpern.

Paul Andersen
Founder and Executive Director, Huts For Vets

Introduction

I HAVE BEEN INVOLVED IN the psychological treatment of veterans for many decades, beginning when I was a neuropsychiatric technician at Valley Forge Army Hospital treating returning Korean vets. But during my time in the service, and throughout most of my clinical career, I, like my mental health colleagues, have been unsuccessful in offering much help to wounded warriors. Then in 2013 I was asked to help found a unique wilderness program for vets (see Chapter Four), and things changed.

This book is aimed at alleviating the frustrations and pains of three groups of readers: It is for vets who need to find ways to overcome their life-damaging war wounds. It is for the vets' loved ones, who need to understand the sources of the vets' sufferings, as well as ways to gain them help. And it is for mental health professionals (MHPs), who are so desperately seeking effective treatment tools for the vet populations.

The unalleviated suffering of vets is well-known, thanks to the media's regular reports on the massive failings of the VA, and the many stories of suicides or crimes and the life tragedies of depressed, angry, non-functioning vets. Not so well reported are the plights of wives and children of returning vets who find themselves trapped with a husband, wife, or parent who is so damaged as to be a liability to his family, community, and himself. And for mental health professionals, there is the frustration of seeing clients getting worse or giving up on seeking help. The opportunity for mental health workers to have effective treatments for vets clearly offers major benefits to the vets, their families, and the community at large.

My preparation for the vet wilderness project at first led me to review the professional literature on treating PTSD. That proved unenlightening, as it involved using the same treatment modalities that I had been employing with marginal success.[1]

---

1. An outstanding exception to the failure of most professional literature for understanding and offering therapeutic directions for returning veterans is the work of Edward Tick, Ph.D. His two books *War and the Soul* and *Warrior's Return* offer important understanding of the world of veteran issues, including ways to help.

I then turned to books written by returning vets. Those books were extremely successful in providing important insights into the reasons for the multiple and unique plights of veterans. However, they offered no direct help for finding effective treatments for returning vets.

My understanding of how to help vets only began to grow during those first three years of being the attending psychologist on multiple four-day vet wilderness trips. During those trips, I witnessed the vets being able to open up and ventilate stories and emotions that had been deeply buried. Such emotionally beneficial outpourings had very rarely occurred in their lives or in any traditional therapy.

And that was only one aspect of what was to be learned regarding the unique healing power of vets' peers. There are a multitude of benefits for vets for spending time with others who not only share their horrific experiences but, more importantly, can also identify with their having evolved into people so alien to their pre-military selves. Vets share a culture, a set of values, and even a language that they understand to be completely foreign and absolutely unlearnable to civilians. And like foreigners from a very different culture, they feel it impossible to communicate complex ideas and reactions to someone who doesn't speak their language, share their beliefs, or know what they have experienced. In fact they expect that trying to authentically communicate who they are would only lead to very negative judgments about behaviors they've exhibited or are engaging in that are so antithetical to American civilian standards.

But those wilderness trips, which serve ten to twelve vets at a time, cannot be the primary model for helping the hundreds of thousands of returning afflicted vets. All this made it clear that the task was to develop treatment techniques based on peer interactions that didn't require a long weekend wilderness experience. It is the high success rate of these new peer-oriented approaches, with their potential for reaching massive numbers of vets, that provided the motivation for this book.

Chapter Overviews

Chapter One: From Civilian to Veteran:
How Vets Got "That Way"

This chapter describes the transformation from patriotic recruit to disciplined soldier to combat warrior to disabled veteran. It reviews the dynamics of the physical, psychological, and moral wounding suffered by vets that handicap them from returning as the leading, contributing citizens their training can potentially foster.

Chapter Two: The Wrong Way to Treat Vets:
Commonly Used Techniques That Hinder Vet Healing

The problems and mistakes of the United States Department of Veterans Affairs, formerly the Veterans Association (VA), as well as other large government and university agencies, are documented and listed as avoidable major errors in the treatment of our vets. The sources of "wrong" diagnostic and treatment methods are also reviewed. The chapter concludes with how the communities' attitudes and practices dramatically affect the adaptation of veterans to civilian life.

Chapter Three: The Right Way to Treat Vets:
Specific Guides for Enhancing the Lives of Vets

The right ways to treat vets are presented as highly veteran-specific focused diagnostic and treatment methods. Highlighted are the findings and experiences regarding the unprecedented values of incorporating peers into various treatment programs. Again, the chapter ends with a focus on the role of communities; here, it is a recounting of how successful transformation into civilian life has been so well implemented through certain community programs.

Chapter Four: Huts for Vets:
History and Procedures of a Successful Wilderness Program

The history and procedures of Huts for Vets are offered as an example of the exceptional value of wilderness for returning veterans. The step-by-step description of the weekend experience is amplified with stories from the trips that illustrate their value. Again, community participation in and support of the program is provided as another example of the critical role of community.

Chapter Five: Three Peer-Oriented Treatment Modalities:
Harnessing the Power of Vets to Heal

This chapter describes how soldiers learn the value of being on a focused mission and never deserting a wounded comrade. These two deeply held values underlie the veterans' high motivation to serve distressed fellow veterans. Add to this the fact that veterans believe only those who have shared their war experiences can understand them and you have the basis for the unprecedented findings of the power of peer interaction for healing. Presented here are three peer-based therapy approaches that can be easily and inexpensively implemented to serve returning vets.

Chapter Six: Peer Mental Health Professionals:
How to Educate Vets as MHPs for Vets

The ultimate professional therapist for vets would be a vet who is, him or herself, a mental health professional. This chapter reviews the advantages of eliminating the middle man, and of the various ways vets can acquire the credentials that allow for licensure as MHPs. Furthermore, this chapter describes the most practical and ideal curriculum that a vet should study in order to meet the psychological needs of other vets. The education of a vet to be an MHP is quite different from that suggested in Chapter Five for non-vet MHPs who are involved in vet services.

Chapter Seven: Conclusions and Suggestions:
Specific Help Suggestions for Vets, Friends and Family, and MHPs

This chapter pulls from the contents of each preceding chapter to offer a collection of specific and practical suggestions for the three groups of targeted readers: veterans, family and friends of vets, and mental health professionals. Each group can benefit from all three groups of suggestions. Vets, for example, can learn from the suggestions for what expertise MHPs should gain before professional interactions with vets, while family and friends can learn from the suggestions for vets what they might encourage in their loved veterans.

Chapter One

FROM CIVILIAN TO VETERAN
How Vets Got "That Way"

THIS CHAPTER DESCRIBES THE TRANSFORMATION of a high percentage of our military recruits from citizens into functioning soldiers, and then into non-functioning vets. It is this troubling end of the story that this book aims to address and impact.

FROM CIVILIAN TO SOLDIER

The military has developed powerful training techniques for turning ordinary citizens into highly functioning "GIs". This stands for "Government Issue" entities, meaning they are first stripped of much of their individuality—given the same haircut, same uniforms, same bed and board, the same rules and customs, routines and purpose. Then these homogenized GIs are taught how to be efficient killers. That involves learning not only weaponry and teamwork but also a level of discipline designed to ensure the following of orders immediately and without question.

I'll not soon forget a basic training exercise that involved me laying down a line of fire between two designated points. The memorable part of that was learning that I could not fire back at someone outside my designated line of fire, even if they were shooting directly at me. I was taught the discipline to maintain my assigned fire zone, to behave as told even in moments of danger and rely on others to do the same. It was up to the person in charge of the operation, not me, to spot and direct fire toward my adversary.

Becoming that kind of an automaton is not necessarily a bad thing. There is pride in being part of a dedicated team. There is the forging of brotherhood or sisterhood relationships that rival the closeness of family or friends. There is self-respect built on the knowledge that one is defending one's loved ones and nation. The tendency to see oneself as a heroic figure commanding the respect and gratitude of one's country is far from negative. Yet severe negativity about oneself and one's country and its policies typify a very high percentage of our returning warriors. Somehow they have transformed from proud, well-adjusted soldiers into veterans who are detrimental to themselves as well as to society. The goal, obviously, is to restore those souls so they can rejoin, participate in, and even serve as role models and leaders for the country they defended.

Potential for Veterans

Veterans have amazing potential for serving in leadership roles as post-service citizens. This is because the military has also taught them how to focus on a mission, how to work as a team. Soldiers are trained to be both superior leaders and followers. The military involves levels of loyalty and wisdom that make for outstanding leaders. Consider the immediate reactions of vets at the scene of the Boston Marathon bombings: Most civilians, acting with understandable motives of self-preservation, ran away from the explosions. In contrast, many servicemen and vets ran *toward* the explosions. They were trained to deal with and help in highly dangerous situations.

Our country needs leaders with that kind of courage, dedication, and self-sacrifice. The problem is that so many vets do not return to civilian lives able to make contributions to society. Too often it is the reverse. Vets return with a variety of troubled ways of being including low self-esteem, anger, inability to focus, deep depression, and substance-abuse problems. They function very poorly as husbands, wives, parents, employees, or even simply as friends. One wife described to me her experience with her veteran spouse with these words: "I no longer have a husband. I now have an extra and very troubled child."

How prevalent are the functional incapacities of returning vets? The statistics indicate very high numbers. For instance, the U.S. Department of Housing and Urban Development (HUD) estimates that 47,725 veterans are homeless on any given night. In 2007, the Bureau of Justice Statistics estimated that 140,000 veterans were in state and federal prisons, and that more than a third had maximum sentences of at least twenty years. Six out of ten of these incarcerated veterans had received an honorable discharge. Then there is the shocking suicide rate among returning vets—according to a study released by the U.S. Department of Veterans Affairs, twenty-two suicides each day.

From Soldier to Veteran

So how does one morph from a dedicated, healthy soldier into a disabled, disenchanted veteran? The answer can be gleaned from the following multitude of factors, which combine in various ways to produce so many poorly functioning veterans.

The story begins with the attitude of those civilians volunteering for the service. The most common reason for joining the service is pure patriotism.

Witness the flood of young people who presented themselves for recruitment following the events of 9/11. Their motivation was, clearly, to serve their country and avenge the deaths at the Twin Towers. Yes, there were also those who join the service for just room and board or for the right to shoot people, but such non-patriotic reasons make up only a fraction of new recruits.

> *"This nation will remain the land of the free only so long as it is the home of the brave."*
>
> Elmer Davis

That idealistic view of soldiering and war is usually maintained until war is actually experienced. That new and very cruel reality can be caused by a number of factors. Here are the major ones.

Disenchantment with Leaders

A terrible blow to soldiers' devotion to the cause comes with a belief that the officers and/or politicians who manage the war are ill-informed, disingenuous, materialistic, unrealistic, and, at worst, willing to sacrifice soldiers for self-centered reasons. The book *The Outpost* by Jake Tapper, which tells the true story of a 2009 Afghanistan battle, offers a very dramatic example of the reality of the high price paid by soldiers for flagrant incompetence and self-promotion by grossly unfit officers. It is heartbreaking to read about soldiers sent to sure deaths out of some officers' stupidity and the unconscionable motivation to make life-and-death decisions based on making their own service records look good. But it is way beyond heartbreaking to actually be a soldier so victimized. It is completely understandable that such soldiers not only lose their positive views of warfare but also act out in terrifying ways, like murdering their own officers, an act that happens with some regularity according to combat soldiers with warfare experience. This is called *fragging*.

Frag is a verb defined as deliberately killing someone of a superior rank, usually an officer. Though reliable statistics are understandably very hard to come by, Wikipedia's information on Fragging offers that there were some 900 such instances in Viet Nam between 1969 and 1972.

The Realities of War

> *"I hate war as only a soldier who has lived it can, only as one who has seen its brutality, its futility, its stupidity."*
>
> Dwight D. Eisenhower

An essentially universal traumatic disillusionment comes with the consistent exposure to a collection of horrors that few civilians can imagine. I remember one veteran telling of his assignment to gather up the bodies of the dead. He was to place them in bags that frequently proved too small and so required him to dismember some bodies so as to fit the parts into the bags. After days of carrying out this morbid and ghoulish task, he simply became overwhelmed and psychologically collapsed. He could no longer consider his service the noble defense of his country, nor justify the war's horrific tolls.

Just filing away in one's mind that "war is hell" is insufficient. Read and try to grasp what living the following stories would be like: Breaking down a door and firing at all "potentially dangerous" inhabitants only to discover that inaccurate intelligence led you to kill an innocent mother as well as her children. Looking into the desperate eyes of a dying close friend you are holding while his innards spill out over you, and then feeling guilty that he was hit when it could have just as easily been you. Witnessing senseless slaughtering of comrades during what you know to be an ill-conceived and meaningless mission. And remember that all such events are not one-time occurrences but something you are forced to observe or participate in day after day, month after month, through multiple deployments.

PHYSICAL WOUNDING

Then there are the effects of physical wounds. These vary from the very visible injuries, such as losing limbs or sight, to invisible ones, such as poorly or non-functioning body parts, brain injuries, psychological illnesses, or moral wounds. With instances of visible wounding, the positive effects of being considered a hero quickly dissolve with the realization of the real-life limitations that injuries such as blindness, or maiming that confines the vet to a wheelchair, impose.

And the invisible physical wounds, such as brain injuries, are worse. A malfunctioning brain causes such unappealing behaviors as outbursts of anger, sadness, and desperation about reduced intellectual or motor functioning. It is very disheartening to know that others now regard you as damaged, a burden, or just plain dumb, and just as disheartening to feel that way about yourself.

As one brain-injured character from Kate Wenner's play about vets in a military hospital, *Make Sure It's Me*, put it very concisely: "Give me back my brain and you can take any three of my limbs."

SELF-ALIENATION

Another very major and not uncommon disillusionment was first revealed to me by a vet explaining what a blow it was to him when he was very suddenly hit with the idea that rather than being part of a liberating force for good he was "one of the bad guys": "All of a sudden I realized that we were invaders not only destroying a country but killing their civilians." In that moment he went from pride to shame. Discovering that a war is not being fought for ideal American values but for commercial reasons has been a rather common experience, especially among post-9/11 vets. It was amazing how many vets on wilderness trips would express this devastating disillusionment with their government's policies. It reminded me of patients whose major symptoms came when someone they revered, such as a priest or a family member, breached an inherent trust. Such psychological blows can destroy one's view of the world and themselves.

The very worst of the psychological war wounds come not from what some soldiers have seen but from what they themselves have done—things done when they've reached a point of no longer valuing any life, including their own. War leads soldiers to levels of anger, fatigue, and disillusionment so beyond anything in normal life that it generates *behaviors* beyond anything in normal life. Most tragically, it leads to such a distortion of the value of lives that killing for fun becomes not only possible but common. Probably none of those patriotic recruits could believe they would come to value their own and others' lives so cheaply.

These transformations are what Dr. Edward Tick, who has extensively researched, studied, and provided a number of healing experiences for vets, refers to as having *lost one's soul.* It is hard for outsiders to believe the atrocities committed by American soldiers, but it is even harder for the vets to integrate the facts of what they were guilty of committing in the throes of warfare into their concept of themselves. They are left with a degree of self-alienation that disallows the receiving of love they feel they do not deserve.

Certainly the most potent experiences I've had with vets came after a Huts for Vets wilderness experience had forged the group into a band of brothers. It was then that the stories would emerge. The first stories would deal with lighter fare, perhaps evaluations of different weapons, funny happenings, or tales of dealing with the VA, but then, invariably, the deeply important stories would come forth. I believe those stories came out for two reasons.

First, the storytellers knew they were with people who had either participated or at least been tempted to participate in similar soul-destroying behavior.

They knew their fellow vets not only would not judge them, but also would, in fact, applaud them for telling the kinds of truths they themselves were keeping inside. The telling of such long-buried stories frequently paved the way for others in the group to give voice to their own sorrowful tales.

And, of course, the second reason these stories came out was a deep need to "confess" and receive understanding for what is usually so deeply hidden. Veterans need absolution. Veterans need to believe that the horrible things they have done are not reflections of their true essences and values but rather the results of being in a war situation that fosters a bestial way of being.

The powerful healing power of gaining some release from guilt and shame that I witnessed again and again during the wilderness trips is what led me to search for ways to harness such healing effects into therapies that could be utilized with vets not in such unique wilderness programs as Huts for Vets. Those techniques are described in Chapter Three, "The Right Way to Treat Vets."

Vets' Response to Vets

I can't leave the subject of vets' disastrously altered self-images without noting the seemingly bizarre responses from other vets upon hearing their fellow vets tell their "confession" stories.

My first exposure to vets responding to vets came early in my experience with vets in the wilderness. At the end of a long day, we were all seated around a crackling campfire. Following a slew of service-related tales of various army screwups, one vet took center stage and told a story so horrific that I found myself actually shivering. The details involved a collection of snafus that ended in this soldier slaughtering a group of innocent civilians. Then even more startling than the story was the reaction of the listening vets. They were laughing uproariously.

I was stunned. How could one find humor in the killing of innocents? Well, they didn't. They understood the horror only too well. That was just the beginning of my learning that some events overwhelm, distort, and alter ordinary responses. What I had witnessed was a strange relative of gallows humor. And it was completely understood by the storytelling vet. He laughed along with the others, feeling bathed in acceptance and understanding. Since that early experience, I've seen it replicated repeatedly. It is not always laughter, but sometimes responses ranging from quiet nods to loud cheers or hoots with verbalizations such as "Yeah man," all of which communicate complete comprehension and approval.

So I've learned that war can lead highly civilized humans to change and to commit crimes so alien to their essences that they become hated strangers to themselves. Many of the dysfunctional behaviors of our returning warriors (suicide, anger, hermitizing, the inability to relate to loved ones or common civilian tasks) are the direct result of the soul-destroying actions, fostered by extraordinary circumstances, in which a sizable number of soldiers have engaged. There are no pills to heal such soul wounds!

PSYCHOLOGICAL WOUNDING

The non-physical wounding, which is the primary injury suffered by a majority of our returning warriors, is what is currently referred to as post-traumatic stress disorder (PTSD). The "trauma(s)" that cause this stress disorder include all of the events referenced above and many more. It is these multiple traumas that morph the typical ideal GI into a troubled veteran. PTSD is actually a syndrome; that is, it can be comprised of a collection of life-destroying symptoms.

It is important to note that it is not only combat troops that suffer traumas. Though combat is the most potent cause of PTSD, and as so has been used in most of the examples offered of service-related traumas, there is also a collection of non-combat traumas suffered by the military. There are the sexual traumas, mostly experienced by our female soldiers. There are traumas brought on by harsh and difficult training events. For a certain number of recruits it is traumatic to realize that they no longer control any major aspect of their lives, including placing their lives in mortal danger. Many who sign up for the Marines or Special Forces are broken by the "right-stuff" filtering methods utilized in their rigorous training. There are traumas associated with many forms of bullying, frequently for sexual or religious orientations. Some soldiers experience being traumatically harassed for not measuring up to some military standard, such as by having an atypical appearance (overweight, tall, short, thin). And, of course, there are the traumas that are generated from the many non-combat service accidents that kill and maim.

To begin to understand post-traumatic stress, review the following description of PTSD symptoms adapted from the Mayo Clinic website. As you review each symptom, allow yourself to empathetically feel what it would be like to endure that symptom year after year after year. And remember, our soldiers have rarely become afflicted with PTSD from one event. Most often there have been multiple traumatic events that significantly intensify the condition.

According to the Mayo Clinic, "PTSD symptoms are generally grouped into four types: intrusive memories, avoidance, negative changes in thinking and mood, or changes in emotional reactions." The following explanation of symptoms comes directly from the Mayo Clinic's website.

Intrusive memories symptoms may include:

- Recurrent, unwanted distressing memories of the traumatic event
- Reliving the traumatic event as if it were happening again (flashbacks)
- Upsetting dreams about the traumatic event

- Severe emotional distress or physical reactions to something that reminds you of the event

Avoidance symptoms may include:

- Trying to avoid thinking or talking about the traumatic event
- Avoiding places, activities or people that remind you of the traumatic event

Negative changes in thinking and mood may include:

- Negative feelings about yourself or other people
- Inability to experience positive emotions
- Feeling emotionally numb
- Lack of interest in activities you once enjoyed
- Hopelessness about the future
- Memory problems, including not remembering important aspects of the traumatic event
- Difficulty maintaining close relationships

Changes in emotional reactions (also called arousal symptoms) may include:

- Irritability, angry outbursts or aggressive behavior
- Always being on guard for danger
- Overwhelming guilt or shame
- Self-destructive behavior, such as drinking too much or driving too fast
- Trouble concentrating
- Trouble sleeping
- Being easily startled or frightened

Dr. Edward Tick writes that military-induced PTSD is not a mental disorder acquired by vulnerable soldiers but rather a condition expected from most of those who participate in war, i.e. a noble and inevitable wound. When referring to the condition for vets he drops the "D" for *disorder* and uses the label PTS (post-traumatic stress) instead. I agree with Dr. Tick's thinking and so use PTSD only for single-event traumas, as is usually the case with civilians but not with vets, where the traumas are multiple and tend to be ongoing over a long period of time.

This book deals primarily with post-9/11 combat vets. That includes women. Women have participated in all aspects of the military, including combat (though that wasn't officially sanctioned until December 2015). Their service experiences, however, have been sufficiently different enough that in programs like Huts for Vets, they participate in separate trips from those with male veterans. Those trips made it clear that women tell different war stories than men. Women mostly have not been in active combat but rather have acted as mission support at forward operating bases (FOBs), where the traumas of war are pronounced and visceral. Women soldiers' second-most-common topics involve the horrors they saw and experienced from those horrific mission-support positions. Their stories from these experiences match both the intensity and effect of male veterans' stories.

> *"If you had seventy near-fatal car accidents in one year of your life between the ages of eighteen and nineteen, do you think that would mess you up all by itself? That's what it was like in Vietnam."*
>
> Jack, Combat Reconnaissance Sergeant

Women's first-most-common war stories involve reports of trauma from sexual intimidation, abuse, and rape. According to the National Center for PTSD, the rate of military sexual trauma for female soldiers is about twenty times greater than that for male soldiers. Women's deployments are often characterized by sexual trauma, which, so sadly, is within the "acceptable" bounds of a too-common fraternal brotherhood of male soldiers by whom women are viewed as legitimate sexual targets. (This pattern unfortunately is reminiscent of the extremist Muslim view of women.) Thus, the mission for the prevention of service trauma for women needs to include the re-education of many male soldiers.

Though HFV has successfully run heterogeneous wilderness trips, including mixing officers with enlisted men, mixing differing races and creeds, and mixing Marines with Navy and Army vets, they never mix males and females on trips. To gain a sense of the experience of women vets, I recommend seeking out their own voices in literature. Helen Benedict's book *Powder: Writing by Women in the Ranks, from Vietnam to Iraq* offers poems, stories, and essays by nineteen women vets. That book is rich in its ability to teach about the nobility and courage that define our women veterans. Additional books excellently documenting and describing various female soldiers' experiences are: Helen Benedict's "The Lonely Soldiers: The Private War of Women", Tanya Biank's "Undaunted", Helen Thorpe's "Soldier Girls", and two by Kirsten Holmstedt, "The Girls Come Marching Home" and the award-winning "Band of Sisters; American Women at War in Iraq."

Conclusions

The transition from being a functioning member of the armed forces to a dysfunctional veteran is brought about by a collection of major traumas. These major traumas, many of which have been described in this chapter, are great in number, extremely common, and occur almost every day in combat zones but are also present in many non-combat military situations.

The next chapters will discuss the wrong and right ways to treat our disabled veterans. For if we treat them correctly and begin to repay our moral debt to those who have served, we will be rewarded with a collection of talented citizens with wisdom and knowledge available for the betterment of our nation.

Notes

You can contact Dr. Alpern with questions or comments at www.VetsForVets.info

Chapter Two

THE WRONG WAY TO TREAT VETS
Commonly Used Techniques That Hinder Vet Healing

On each Huts for Vets trip I ask the group to discuss their experiences with mental health professionals (MHPs). Alarmingly, vets considered less than 10% of their VA mental health contacts to be helpful. I heard story after story about contacts that offered no help, made things worse or much worse, and certainly discouraged the vets from seeking any further professional help.

VA Treatment Stories

The typical story vets tell about the treatments offered at the VA goes like this: The vet finds, after filling out multiple repetitive and confusing forms, followed by a long wait, that their application has been lost. After redoing the noxious process and waiting some more, the vet is awarded, if they're lucky, with an actual appointment—albeit for months in the future. After several more months of waiting, the vet finally has their appointment at a VA facility. They meet in an institutional building with a young therapist who not only seems new to their profession but has also never been in the service (let alone combat), and has no knowledge of the complex military culture. The vet would have a difficult enough time explaining their dilemmas in some way understandable to a civilian, but even that fails to be a possibility. Rather, the "professional," while looking primarily at their computer, asks a collection of predetermined questions such as "How are you sleeping?" "Have you been contemplating suicide?" "How are things going at your job? With your family?" After forty minutes the appointment ends with the vet being given prescriptions for all "symptoms," which the professional dutifully records.

In addition to the personal testimonials of vets there is impressive hard data testifying to the failures of the military and the VA's treatment of veterans' PTSD. In the August 2015 *Journal of the American Medical Association* (JAMA), one article, "Psychotherapy for Military-Related PTSD," reviewed thirty-six studies reporting on treatment of 2,521 veterans. It concluded, "PTSD treatments have high non-response and dropout rates and patients remain symptomatic." That article offered the conservative conclusion that there is a need for improvements in existing PTSD treatment. In the same issue of JAMA, David J. Kearney and Tracy L. Simpson stated in their editorial article, "It is essential to develop therapeutic alternatives for patients." And such ineffectual VA treatments come at an astronomical financial cost: for example, between 2007 and 2010 the Department of Defense spent over 2.7 *billion* dollars for treatment and research on PTSD. And in their June 2014 final assessment report on treatment for post-traumatic stress disorder, the Institute of Medicine concluded, "a lack of standards, reporting, and evaluation significantly compromises" DOD and VA efforts. The departments often do not know what treatments patients receive or whether treatments are evidence-based, delivered by trained providers, cost-effective, or successful in improving PTSD symptoms. The departments also collect little information about the effectiveness of their programs in the short or long term.

Further documentation of the dysfunctionality of the VA was reported in *USA Today*'s 11/11/15 front-page story "VA Bonuses Paid Amid Scandals." It revealed that $142 million was paid out in bonuses to executives and employees for performance in 2014 even as scandals emerged describing their despicable negligence and corruption. For example, the executives who oversaw a years-overdue construction project in Denver—which was also over budget by more than one billion dollars—took home $4,000 to $8,000 each in performance rewards. And in 2013, the VA paid more than $380,000 in performance bonuses to top officials at hospitals where the veterans' unconscionably long wait times for treatment reflected the poorest management—and these top officials included some administrators under investigation for wait-time manipulation. That is, they repeatedly lied about how long vets had to wait to be seen. And on March 30th 1916 the title of David Phillip's article in the New York Times "Report Finds Sharp Increase in Denied VA Benefits" tells the updated story.

Some of the stories vets told of their VA contacts involved being assigned, usually in addition to their collection of drugs, to a certain therapy. The VA utilizes treatments with known validity, i.e. empirically validated treatments, for the admirable reason that they are backed by good research data.

> *"There is clear evidence from internal investigations in the past that some raters actually see themselves as adversaries to veterans. If a claim can be minimized, then the government has saved money, regardless of the need of the veteran. Just recently, the press exposed an official e-mail from a high-level staff person who stated in essence that PTSD diagnosis was becoming too prevalent and offered ways to delay and deflect ratings in order to save the government money."*
>
> TAYLOR ARMSTRONG,
> HIDING FROM REALITY

Most often used are prolonged exposure therapy, cognitive therapy, or eye movement desensitization and reprocessing (EMDR). The trouble is that these treatments have been validated as successful primarily with *civilian* populations. Too often the treatments are not appropriate for the specific kinds of traumas suffered by vets, nor ones to which vets respond. For example, cognitive therapy involves the client describing his or her thought processes. But the veteran population is well-known for, even characterized by, their strong prohibitions from describing their service experiences or thoughts.

Need anyone explain why such appointments are not helpful?

The VA is a huge bureaucratic organization that has failed veterans in many ways. The interminably long waits for help have become legendary from newscasts displaying the large rooms stacked with overspilling piles of waiting veterans' files. Then a possible fix was announced, a new program wherein if the vets had to wait for more than sixty days for a VA appointment they were authorized to seek help from local private facilities. Around a Huts for Vets campfire one vet brought down the house with his not-so-funny tale of waiting over ninety days to have his application approved for the program, thus taking longer to apply for the hurry-up program than for his actual VA appointment.

But delayed help is far from the worst mistreatment vets can suffer through the VA. Here are some of the major problems: First, the VA's huge budget is protected through the unsavory practice of intentionally misdiagnosing. This is commonly because pills are far cheaper to dispense than professional treatment time. A diagnosis of PTSD costs much, much less to treat than traumatic brain injury (TBI). There are doctors who have lost their jobs with the VA for providing diagnosis of brain injury. Furthermore, there are credible stories about VA staff informing veterans and their families, incredibly, that no TBI condition exists. Huts for Vets produced a play that tells that exact story. That play startled a former VA physician, Dr. Chrisanne Gordon, who said that it exactly reflected her own circumstances, having been fired by the VA for "over-diagnosing" brain injury. Dr. Gordon now runs a *private* clinic treating brain-injured veterans where she feels she is not prevented from diagnosing and treating veterans in accordance with her expertise and not limited by bureaucratic rules which violate her medical professional standards and ethics.

The second major problem with the VA's psychological treatment for veterans involves the nature of usually long-awaited appointment with an MHP. Inside an institutional office, a usually new and inexperienced (i.e. cheaper) therapist sits in front of a computer screen and fires the collection of preplanned, one-size-fits-all questions at the vet. Then after forty minutes the vet is dismissed with new or refill prescriptions and a follow-up appointment in the distant future. That is not a burlesquing of the typical VA mental health appointment. If anything, it downplays what so many vets have described, a cold appointment where they were essentially never looked at, did not get help in talking about their situations, and were not motivated to keep the distantly scheduled follow-up appointments. During the Huts for Vets trips I always ask the vets about their VA mental health experiences and yes, I do hear tales of very positive help gained at the VA—always involving that rare experienced and compassionate therapist whom they respected and found helpful. But for every one of those there are ten or fifteen that emulate the above negative story.

A third failure of VA mental health treatments is one unfortunately shared by the universities training MHPs to work with traumatized vets. That problem, mentioned above, is the use only of treatments that have been empirically validated. That sounds very reasonable; however, many of those treatments were empirically validated on populations of patients very different from veterans—namely, college sophomores or civilian patients who have not been indoctrinated by the armed forces, and so have not been thoroughly infused with a very specific set of values, ways of being, approved reactions, and reactive emotional habits accrued from combat duties. The use of treatments standardized on one population does not translate to a population from a very different culture. There are a number of treatments that have been creatively designed especially for the vet population that have had impressive success and, incidentally, are generally much less expensive than those delivered by the VA. But these effective treatments are not condoned, nor taught or used, within the VA or most university clinics.

A final deadly failure by the large government entities that treat vets is that they don't properly utilize a major healing modality...veteran peers. Vets will voice, and frequently shout the sentiment: "If you weren't there you can't possibly understand." Vets don't trust those who they believe can't understand what they went through, are liable to judge them for actions civilians consider reprehensible, and have never heard of, much less have skills in treating, "soul" or "moral" wounds. When working with therapists who you know can't understand you and might judge you, you can't be successful. But who does understand vets, especially combat vets? The answer is vets who have similar experiences. Not using this resource is not only denying the vets therapy they can trust but also denying the helping vets a mission that they covet and can do better at than most professional mental health workers.

Inaccurate Diagnosis

A basic tenet of any treatment is that it be preceded by a good diagnosis. A major problem with VA treatments is that they are often administered without an appropriate diagnostic evaluation. Rather, the VA tendency is to simply gain a list of symptoms and then prescribe a drug for each symptom. The average vet ends up with a collection of five to ten drugs. And, remember, giving only pills costs the government just pennies.

Worse than symptom treatment sans an adequate diagnostic evaluation is, and this is critical, how the VA actively discourages certain diagnoses. How could this be? Why would the VA discourage or even prohibit certain diagnoses? The answer is money. The major example is the intentional under-diagnosis of brain injury, which is extremely common with post-9/11 vets because of their so-common suffering from repeated concussions from improvised explosive devices (IEDs). A diagnosis of brain injury requires months and possibly years of expensive treatment with specialists. On the other hand, most VA treatments for PTSD involve a matter of weeks.

Therefore the VA prefers, and even mandates, that brain-injury symptoms be attributed to PTSD.

Physicians were even dismissed from VA clinics and hospitals for "over-diagnosing" brain injury, including the aforementioned Dr. Chrisanne Gordon. Dr. Gordon was working for the VA for very little compensation, essentially volunteering, out of her desire to serve our veterans. Although Dr. Gordon has impressive expertise in brain injury, she was fired by the VA for "over-diagnosing" it. Dr. Gordon continues to treat brain-injured vets, but now at her own private clinic funded by her own and donated private funds.

It was Dr. Gordon who introduced me to Kate Wenner's play *Make Sure It's Me*, stating that it so accurately reflected her real-life story. The play takes place in a military hospital and depicts the plight of returning wounded vets and their families. It deals with vets being denied appropriate treatment by the military's administrative policies.

Huts for Vets then produced that play, using vets from our trips as the actors. One of the most dramatic aspects of the performances was following the play when the performing vets stayed on stage for a Q&A. The real vets conveyed to our audiences how the play's plot mirrored their real-life experiences with VA treatments.

Military Medicine Protocols

The overuse of treating symptoms with drugs for financial reasons is but one example of the military's non-patient orientation. Drugs are also used for reasons completely antithetical to patients' well-being. One would assume that a battlefield physician would be primarily dedicated to the health and well-being of the troops. Not so. A major task for battlefield physicians is to keep the troops on the line fighting. So, for example, a soldier who reports that he is overwhelmed with stress and not able to function in his battlefield job may well be given medications designed to return him to battle. Anti-seizure or anti-psychotic medications have been reported as the pills of choice to get the soldiers back on the line. That is a travesty of the Hippocratic oath to "first do no harm."

Military medicine has long had the policy that physicians compromise patient needs for military goals. For instance, in World War One when soldiers incapacitated by trauma refused to get back in the trenches they were frequently given electric shock. This particular electric shock was not the kind currently used to treat unrelenting depression, which is essentially without physical pain. The electric shock administered during World War One was designed to be as painful as possible so that the soldier would choose to go back to the front lines rather than be repeatedly tortured.

> *"The responsible choice would be to honor those who have worn our nation's uniform, but the administration made a different choice. They're raising veterans' health care fees by $250 a year while cutting taxes for millionaires."*
>
> SENATOR JOHN KERRY

In addition to the cruel and non-patient-oriented treatments, there are a myriad of ineffectual treatments used by the military. My experience as a Korean War neuropsychiatric technician illustrates this point. I was stationed at the Valley Forge Army Hospital from 1954 to 1956. There I was wardmaster on a unit administering insulin shock therapy to Korean and World War Two vets. What we did was inject our vet patients with massive amounts of insulin designed to cause them to go into a shock that was, theoretically, intended to help them. Back then I believed I was doing a good thing. In fact, insulin shock therapy had years before been researched and reported on in the well-known medical journal *The Lancet* as "ineffectual and inhumane." So the military was using an archaic treatment method that tortured the vets without having any benefits. How many other inappropriate treatments have been and are being used that offer no help but cause various degrees of discomfort and, in fact, make things worse? The most common of these come with the strong negative effects of over-medicating.

Then there is the problem of the VA not having the staff or funds to permit timely appointments for vets. Timely (i.e. early) treatment is critical for many vet-related conditions. Months- and year-long waiting lists for VA appointments is a national scandal. Delayed treatment is the equivalent of denied treatment.

And I have to reiterate the fact that the VA mandates the use of treatments that are sanctioned by their central command office. The treatments that are so approved are those that have been "empirically validated." But such treatments have typically been validated with college students or civilian mental health patients, *not* with the veteran populations. The nature of war traumas, coupled with the unique culture of vets, renders them a separate population needing specialized trauma treatments, including specialized mental health practitioners. Another problem with centrally dictated treatment modalities is that they lack sufficient flexibility to fit the radically differing individual traumas suffered by vets.

Perhaps the most grievous of the military techniques for dealing with our wounded vets is the inappropriate over-reliance on drugs. Certainly the most frequent VA "treatment" consists of a multitude of pills. Pills are so universally used as they cost only pennies and do not require therapists of any kind. Pills are also problematic as they are primarily used by the military to treat only symptoms. Vets repeatedly tell how the massive numbers of VA-issued drugs have damaged them so badly that they consider their first step toward recovery being to get rid of their deadening drugs with their multitude of debilitating side effects. Additionally, if the vets refuse to take the pills they are required to sign off on refusing medical treatment, which limits their access to various earned vet benefits.

Improper Treatment Organizations

The fact that military and veteran government organizations are prone to the devastating errors listed above is why I believe the treatment of vets is best undertaken by private profit and non-profit organizations. The government system, by either design or ignorance, is unable to successfully treat a significant portion of our active soldiers or wounded warriors. By contrast, there are a number of non-government agencies that have proved very successful in rehabilitating veterans. Those successful methods are reviewed in the following chapter.

In addition to the financial and military goals that account for the government's failure to provide effective treatment to soldiers and vets, there are also explanations related to size. This is a case where the government programs (unlike banks considered too big to fail) are too big to succeed. A huge organization like the VA tends to disallow flexible adaptations, as there are centrally mandated policies that dictate what treatment modalities can be used. Furthermore, the VA requires the use of traditionally university-trained MHPs who learn treatments that are most easily researched. The following illustrates and documents these fatal flaws.

Improper Treatment

How is it that treatments that don't work and even make things worse remain the "treatments of choice"? An example involving a now-traditional, widespread treatment for trauma illustrates the point.

In his book *Redirect: The Surprising New Science of Psychological Change*, Dr. Timothy Wilson describes how federal, state, and even local governments consistently utilize critical incident stress debriefing (CISD) in spite of the fact that there is very good research indicating it is not only ineffectual but damaging.

CISD is used when people have experienced traumatic events such as mass shootings, natural catastrophes, or terrorist attacks. The system involves sending in large numbers of trauma counselors to help victims air their reactions as soon as possible under the theoretical belief that bottling up feelings leads to more serious post-traumatic stress. CISD sessions last a number of hours in which victims are asked to describe the traumatic event, express feelings and thoughts about it, and verbalize any psychological or physical symptoms they are experiencing. That seems a very logical approach. Our cultural belief is that getting people to talk about their feelings is much healthier than suppressing them. Except that careful research has proven that to be dead wrong. Expressing thoughts and feelings about traumatic events immediately after their occurence in CISD tends to *cause* psychological problems. Research with people who underwent CISD right after a trauma demonstrated that it impedes the natural healing process and can "freeze" memories and negative emotions related to the event.

In one very careful study, groups of people who had been severely burned were randomly assigned to either receive CISD or not. Thirteen months after the event, the CISD groups had a significantly higher incidence of post-traumatic stress, were more anxious and depressed, and were less content with their lives.

Those results have been so replicated in a series of similar CISD studies that in 2003 an international panel of renowned researchers selected by the American Psychological Society recommended in the journal *Psychological Science in the Public Interest* that "for scientific and ethical reasons professionals should cease compulsory debriefing of trauma-exposed people." Yet even today, the most typical government response to public traumatic events is to send in the counselors to do CISD.

Dr. Wilson's book offers other research-backed treatments that, in fact, do work extremely well for treating civilian trauma victims. For example, it suggests using the treatment outlined in James Pennebaker's book *Writing to Heal.* But those too-large-to-succeed government agencies continue to use the unsuccessful CISD. *Redirect* also offers successful treatments for traumatized vets far superior to those in unrelenting use by the large service and VA agencies. It's very unlikely that large ossified government entities will ever adopt the newer, more successful treatments. However, the private sector has begun to offer such treatments, and that is where hope lies.

Inappropriate Training of Mental Health Professionals

Veterans' treatments are generally either done or directed by trained MHPs. One problem with this is that the universities that are training trauma professionals are prone to the following collection of errors.

Universities have long suffered from a hardening of their theories. Innovative scientists have a history of leaving academia in order to explore non-traditional ideas that have proved phenomenal. From Isaac Newton to Bill Gates, there is an impressive list of successful innovators who felt the need to get away from universities to pursue ideas considered antithetical to prevailing thought. And then there is the special case of universities' attitude toward academic psychology.

The university powers that be have long considered psychology a "soft" and thus second-class science. If a theory or treatment modality could not be demonstrated to be valid using empirical methodology, then it garnered little respect (or money for grants, faculty, or even space in the university buildings or curriculum). Freud, Jung, and even more modern psychological theorists have not been traditionally amenable to empirical verification and so have been thought to be pseudo-scientific or, worse, labeled junk science. Academic psychology was then motivated to move toward those theories or practices that could be measured by the tools available. Today there is a collection of trauma treatments that lend themselves and thus have been empirically demonstrated to be useful. But again, the trouble is that the available research on these treatment modalities has been primarily accomplished with *civilian* populations, not veteran populations. The differences between combat veterans and the typical college student (a very commonly used group for research) is about the same as the difference between graduate students and pre-schoolers. The point is that you cannot reach conclusions regarding the effectiveness of treatments for vets by testing them out on college students.

Many years ago a newly immigrated Chinese couple sought me out for marital therapy. It took very little time for me to realize that as I did not understand their culture, their beliefs about marital roles, or, actually, their Asian personality patterns, I could not effectively treat them. Likewise, when I, an ex-Korean soldier, attempted to treat returning combat Vietnam vets, I discovered that because I did not understand their military culture, their beliefs about military experiences I could not even imagine, or, actually, their combat-imposed personality patterns, I could not treat them. The point is that for anyone to be successfully treated they must know that who they are is understood.

Returning vets are all too aware that those who have not experienced the service culture and experiences have no basis for understanding them. That is why so many vets live in silent isolation. They do not believe they can explain who they have become to anyone who has not actually shared their horrific, personality-changing experiences. They do not trust that non-vets can comprehend or even imagine who they are. And they are right.

Many university departments training mental health professionals do not get this. Recently I was asked to review a curriculum for a new graduate program designed to train doctoral students to deal with trauma, primarily with veteran populations. I was not shocked to find that nowhere in the training of these professionals were there any courses or even parts of courses that educated the students on military culture. And the culture of the military is radically different from any civilian culture. To train people to work with vets and not prepare them to understand the culture of the vets is not just unacceptable but flagrantly unethical.

And that's not the sole criticism I had of the university graduate program for training doctoral students to treat veterans. Another problem was the already-mentioned emphasis on treatments that are easily quantified, eliminating those harder to quantify in spite of their success. The curriculum did not, for instance, include wilderness programs, which most any participant would tell you have been much more effective than the typical VA treatment. But the most glaring omission from the reviewed curriculum was the lack of the use of fellow vets in the treatment procedure. Absolutely the most important discovery I and others have made in working with vets is that they primarily only trust those who have experienced what they experienced, and trust is the basic ingredient in any therapeutic relationship. How many times need some version of "If you weren't there you can't possibly understand" be heard? Yet the university curriculum takes no heed of this critical factor. In contrast, this entire book's treatment focus is on the use of fellow warriors to implement healing of war wounding.

Military Cultural Anomalies

To offer a sense of how different military culture is from civilian culture, try to fathom the mindset of combat soldiers who have engaged in the voluntary killing of disliked or distrusted fellow soldiers, usually dangerous officers (fragging). An even less believable event to civilians is the intentional killings of civilians, including children. From a group of vets swapping their deeply hidden confessions I have heard stories of recreational killing. That's right: within the military culture there is a niche in combat zones wherein killing for fun is understandable.

Part of the therapy of wilderness trips involves sitting around an evening fire and exchanging stories. A major factor in such trips is that the vets bond with fellow vets, feel they are finally among understanding "brothers" or "sisters," and can verbalize the horrible things that they've not only seen but, worse, found themselves doing. It was on my initial wilderness trips that I first watched one vet, a former Marine sergeant, visibly shake with emotion as he told of how he repeatedly slaughtered innocent civilians. The story began with how his regular patrols involved gathering up groups of suspected civilians and delivering them back to central command for interrogation. However, early on he discovered that some who were released following interrogation were later discovered to be combatants attempting to kill our troops. He ended the story by simply saying, "I never brought anyone back for interrogation."

Everyone around the fire understood that he simply killed every suspected civilian he encountered on patrol. The universal response of the listening vets went beyond acceptance. He was cheered. I'm relating this particular story as it illustrates the repeated slaughtering of large numbers of civilians for a reason that many veterans, even non-combat vets, may be able to identify with and understand. Understandable or not, that Marine suffered the kind of retrospective massive guilt that has turned so many of our brave warriors into non-functioning, suicidal vets.

Like many cultures, the military uses language different from that known or understood by non-military populations. It's very interesting listening to vets talking to one another, as they use so many acronyms that it's very like listening to someone talking a foreign language. They use these acronyms to describe weaponry, tactics, orders, procedures, and many other things that are beyond understanding by all of us not part of the combat culture. And that's the point. Like parents speaking in some code not understandable to their children, vets can exchange ideas with other vets knowing non-vets will not follow. In addition, that method of exchanging thoughts functions as a test. If you understand, it means you're part of their in-crowd, with similar experiences. If you don't understand, it means you have not shared their unique experiences.

Again, to illustrate the significant differences between the thinking and behavior of vets and non-vets, I recall my surprise at the answer to a question I asked during one wilderness trip. The question to a group of about a dozen veterans was what they thought about the use of drones. Because drones allowed for the destruction of the enemy without risking lives I was fully expecting to hear that they approved of drones. Wrong! To a man they disapproved of the use of drones. My next question, of course, was why. Their answer demonstrates their military-indoctrinated value system. They described drones as an unacceptable, immoral method of killing. One man stated that it was honorable to actually fly a warplane, as it put your life at risk, which is not the case with drone flying. Another explained how he'd felt somewhat ashamed when he and his unit had weapons far superior to the enemy's. The continuing stories of what was to them honorable and moral in war repeatedly demonstrated a value system that differed in major ways from that of most civilians'.

It would take a longer book to describe the differences in the thinking, emotional responses, and values between veterans and non-veterans, but the point is simply that if you are not well acquainted with the military culture, vets cannot and will not trust you to relate to them, much less to treat them. Furthermore, it is unethical for MHPs to offer services to populations they do not and cannot fathom. Yet I witness universities training professionals to work with veterans and the VA hiring therapists with no requirement that they be knowledgeable about the culture of their potential clients.

The Role of the Community

Although the role of the community is presented last in this discussion of the wrong way to treat veterans, it is of huge importance. In his two very significant books *War and the Soul: Healing Our Nation's Veterans From Post-Traumatic Stress Disorder* and *Warrior's Return: Restoring the Soul After War*, Dr. Edward Tick offers undeniable evidence of the critical role of the community in the mental health of veterans.

Here's a prime example: the rate of serious PTSD among U.S. Vietnam veterans was extremely high, about 85%, whereas the rate of PTSD among

Vietnamese soldiers was essentially nonexistent. How could this be? The answer is in how the community to which the vets were returning regarded them. To the Vietnamese population, their returning soldiers were heroes who fought to defend their country from the enemy. But in the U.S., the Vietnam War was so unpopular that those who fought in it have been vilified by a substantial portion of American citizens.

One of our HFV board members, Vietnam vet Dan Glidden, teaches this to various civilian groups with two photographs. The first is the famous scene from Times Square following the announcement of the end of World War Two. Clearly that happy embrace between the returning sailor and welcoming young lady indicated the appreciation and honor bestowed upon those who fought for our country. The second picture depicted Glidden's personal reception on returning from Vietnam. It shows him being greeted by civilians giving him and his fellow returning soldiers the well-known middle-finger salute. He tells of the gesture being accompanied by hoots and hollers of "baby killer!" and similar derogatory insults. Is it any wonder that World War Two vets integrated so well into civilian lives while Vietnam vets so very often have not? And such damaging insults continue to exist for post-9/11 vets. Consider this too-common and unbelievably egregious comment: the vets deserve no special consideration as they *volunteered* for the service. The implication is that those patriotic recruits deserve what they suffer since they volunteered to go to war. As if naïve civilians could possibly predict the horrors they would see, the battering they would receive from repeated deployments, or the soul-destroying negative self-concept they would develop from their own unpredicted inhuman thoughts or behaviors.

An important initial step in helping the community avoid misunderstanding, or mistreating, our returning warriors is simply education. Again, this is best done within the community by private organizations. The chapter on the right way to treat vets offers multiple methods for accomplishing such education. This explanation of what the community does wrong must begin with the simple lack of involvement with our armed forces. We no longer have a draft. Not having our sons and daughters risking their lives results in much less concern, interest, and emotional investment in our current wars. Less than 1% of American families have a relative in our armed forces.

During World War Two, the population was highly motivated to participate in the war effort. Americans from every walk of life were accepting and often very proud to have their children participate in what was considered a noble cause. Citizens were also completely supportive of sacrifices such as rationing gas and food, practicing air raids, collecting scrap metals, growing "victory gardens." Our whole economy changed so that auto companies produced military vehicles, not cars. Women replaced men in war effort factory jobs. In stark contrast, today's citizens are asked to make no such sacrifices. Life has changed little for the vast majority of our citizens throughout our recent and longest and most expensive wars.

It is wrong to keep our citizens unaware of the real financial and human costs of our current wars. It is wrong that most Americans have no idea in how many wars and on how many fronts we are fighting. It is wrong that civilians are not being asked to participate in any real way in our wars or in caring for our vets.

Next: the right, the moral, and the appropriate ways to treat our vets.

Notes

You can contact Dr. Alpern with questions or comments

at www.VetsForVets.info

Chapter Three

THE RIGHT WAY TO TREAT VETS
Specific Guides for Enhancing the Lives of Vets

THE NEWER, MORE INNOVATIVE WAYS to help veterans are not found in the VA nor, unfortunately, in the universities. The successful programs described in this chapter have come from private profit and non-profit organizations. Sometimes these have been created by communities. For example, the Welcome Home program in Montrose, Colorado (www.WelcomeHomeMontrose.org), is a hugely successful community-wide project that has helped and is helping large numbers of veterans make a successful transition into civilian life. Similarly, the Huts for Vets program (www.HutsForVets.org), based in Basalt, Colorado, is a non-profit that not only runs free wilderness programs but also has met many of the "fallen through the cracks" needs for veterans. That program provides such a strong model for non-government, non-university treatment programs that it will be described in separate chapter.

Other sources of effective treatments for veterans can be private business organizations. Enabled Enterprises (www.EnabledEnterprises.com), for example, is a for-profit private organization that sponsors vets and helps them become successful competitors in the free market system. They help to manage the Children of Heroes Foundation as well as a veteran-owned small business program. So even where there is a profit motive, there are innovative ways to substantially help our veterans.

This chapter begins with the importance of customizing treatments to meet very individual veteran needs and then describes treatments that have the flexibility to accomplish just that.

THE NEED FOR INDIVIDUAL DIAGNOSIS

A critical initial point to make about the treatment of vets is that whether you're dealing with treatment from loved ones, professionals, or communities, "one size fits all" treatments are going to fail. Successful treatments need take into account the differences between veterans.

Veterans differ from each other in ways that mandate customizing help on an individual basis. Female vets differ from male vets not only because of gender membership but also because the military treatment and expectations differ for the two sexes. The major differences between combat and non-combat soldiers needs no description. Then there are the differences between the branches of the service, the ranks in the service, and the degrees of physical/psychological handicapping suffered. And there are significant differences associated with each war. Some differences between the veterans from different wars are based on the civilian reaction to a particular war; think of the public's reactions to and participation in the Civil War, as opposed to World War Two or the Vietnam War.

The kinds of psychological and physical wounding differ from war to war. Massive numbers of major drug addictions began during the Vietnam era, whereas brain injuries from repeated IED concussions have been a very common injury only for post-9/11 war vets. And although post-traumatic stress symptoms (by many names) from different wars share a number of commonalities, there are also profound emotional differences based on the fact that the nature of the major traumas differs from war to war. Psychological defenses and injuries from fighting in trenches differ from those of fighting in jungles, both of which are majorly different from urban warfare, which is again different from desert warfare. Then there are the profound differences between fighting an enemy wearing distinguishing uniforms and fighting an enemy of civilian men, women, and children wearing civilian garb. And in the latest wars, there is the new, lethal danger of being killed by someone you could reasonably assume was on your side. I'm referring to troops dressed in the uniforms of an army you're supporting or of civilian police. Having to be wary of the uniformed "comrades in arms" beside you is a new threat that produces a hyper-vigilance unlike that of more traditional warfare.

Another important difference that needs to be understood is the range in kinds of guilt suffered by vets. Collective guilt for destroying a country's infrastructure or culture is very different from individual guilts. Probably the most devastating form of individual guilt is dealing with atrocities you yourself committed. The damage to personalities from witnessing the tragedies of war does not compare to the devastating effects of believing that you yourself have perpetrated barbaric acts. It is not known what percentage of soldiers actually engage in such atrocities as killing for fun, but

I have now heard enough such stories from vets that I know it is not at all a rare event.

Very common are those vets who suffer from inappropriate guilt such as survivor guilt, which from a logical standpoint makes zero sense. The question *Why wasn't I killed when those around me were?* can and frequently does have as devastating an effect as the actual responsibility for inhuman behaviors.

Another form of guilt involves not acting in accordance with the unrealistic hero-soldier expectations with which vets were indoctrinated. So, for example, if one did not risk their life to attempt to save another, even though the chances of either surviving were minimal, there can be guilt that leads to unrelenting self-loathing.

The large number of vets who feel their actions or inactions justify guilt believe themselves to have turned into people that they, as well as their loved ones, can neither recognize nor accept. This self-recrimination is, for many, what underlies the well-known phenomenon of vets refusing to ever talk about their service experiences. In retrospect, it is a good bet that the high, high rate of suicide among vets is based on self-hatred. Dr. Rita Nakashima Brock, director of the Soul Repair Center, and co-author Gabriella Lettini have eloquently described and analyzed these kinds of wounds in their book *Soul Repair: Recovering from Moral Injury After War*.

All these differences mandate differing therapeutic approaches. Soldiers who were traumatized by sexual abuse need treatment that differs substantially from that of those suffering from witnessing war horrors. And those who suffer from self-loathing from any of the guilts described above require still a completely different treatment approach. This is why the following "right ways to treat vets" all involve the flexibility necessary to allow customization for each individual vet in accordance with their individual treatment needs.

THE CRITICAL FACTOR: PEERS

Traditionally, diagnosis is a formal procedure accomplished by professionals. It occurs in offices or clinics using classic diagnostic tools, typically psychological testing or mental status interviews. All such usual diagnostic procedures are badly compromised in the case of our veteran population. Vets classically are unable to talk about either their experiences or their reactions to their experiences, largely for reasons

> *"The victims of PTSD often feel morally tainted by their experiences, unable to recover confidence in their own goodness, trapped in a sort of spiritual solitary confinement, looking back at the rest of the world from beyond the barrier of what happened. They find themselves unable to communicate their condition to those who remained at home, resenting civilians for their blind innocence."*
>
> David Brooks

already described: It is much more than deep shame, or conscious and unconscious denial, that keeps vets from describing their states either verbally or in written tests; it is that they feel it futile to communicate to those who couldn't possibly understand, i.e. those who have not experienced the unspeakable. The only people vets can trust to understand their experiences and reactions are those who shared them and speak the same language…other vets.

During Huts for Vets trips into the wilderness I repeatedly witnessed the same small miracle: Vets who had, prior to the trip, been bottled up ended up pouring out their stories to their peers. They told of the horrors they saw and committed. They told of their struggles with the VA, their families, their jobs, and, especially, themselves. All of these outpourings happened because they were with peers who they knew understood their language, shared their experiences, and were not shocked by the behavior of their fellow warriors, nor negatively judgmental of what they did.

Furthermore, I saw the immense therapeutic value of this peer companionship. There was laughter based on the relief of getting it all out. There was new hope from being understood by intimates who were in differing stages of working through the transitions from warrior to civilian. And the term "intimates" is a very accurate descriptor of the inter-relationship that occurred between the vets on the wilderness trips. It was amazing how over the short period of a weekend such immediate bonding could take place between these strangers—and how this bond lasts among the vets who live in the same vicinity, even encouraging some to form their own vet organization. Therapists know how important trust is for the therapeutic process to work. Developing trust with vets is very hard, not only for the typical therapist but even for loved ones of returning vets. Yet with other vets trust was not only easily established but also considerably deeper than the trust therapists typically develop with long-term patients. The vets would tell each other, *I have your back*. When they said that they were declaring a solemn promise to risk their lives for their truly cherished comrades.

So then the question arose of how to utilize this therapeutic bonding to help not just the small numbers of vets who participate in wilderness programs but also the thousands and thousands of other wounded vets. An answer involves three peer-oriented therapy approaches: peer group mentoring, individual peer mentoring, and peer and individual wilderness therapies. These three treatment modalities are described in Chapter Five.

The Roles of Friends and Family

What are the right ways for friends or family of vets to treat their returning warriors? They are the overlooked victims of the many, many vets who return so damaged. They desperately want back the man or woman who left to serve our country. Wives too often feel they have lost a husband and received in exchange a needy, very troubled child. More than once I have heard the spouses of vets guiltily whisper the unspeakable: "It would have been easier had they been killed." Similarly monumental are the losses suffered by vets' children, parents, and loving friends. What can those victims do toward restoring the psychological health of their returning vet?

An excellent web source of information for family and friends is maintained by the U.S. Department of Veteran Affairs. You need only log on to the site *National Center for PTSD* (ptsd.va.gov). Select the "Public Section" link, then click on the "Family and Friends" tab. There you can find authoritative and very user-friendly information on dozens of topics of particular interest to those who are dealing with returning vets. For instance, if you are a primary caregiver to a vet you will find a link to the VA's Coaching into Care program, where you can get information, guidance, and help with caregiving.

An initial step for anyone dealing with a vet returning from war is to not expect him or her to return unscathed. To anticipate normal functioning from a returning warrior is not only naïve but also a setup for significant disappointment. Rather, prepare for the veteran's homecoming with as much knowledge about their experiences as possible. The same learning necessary for MHPs who treat vets is at least as important for veterans' loved ones. Courses online can teach much about the military culture. Books by vets, some of which are included in a reading list in this book's Appendix A, offer wonderful insights into not only the horrors of war but also the emotional and behavioral changes brought about by war. Participating in community veteran programs so as to meet and be educated by other vets provides opportunities to gain understanding, as well as to potentially meet vets who can offer much help to returning vets. Vets who have struggled through the difficult homecoming process have much guidance to offer newly returning or long-suffering isolated vets. They are the ones most likely to know programs, activities, and groups that can be so helpful to those handicapped or incapacitated by their service.

> *"What cannot be talked about cannot be put to rest. And if it is not, the wounds will fester from generation to generation."*
>
> Bruno Bettelheim

Another step is to encourage the vet to take advantage of the real help that is available. Vets are notoriously disgusted by and hopeless about the unhelpful "help" forced on them by the military both during and after active service. Loved ones knowledgeable about the "wrong" and "right" ways to help vets can guide and encourage the vet toward successful healing. This can require immense patience. Such patience is fortified by acquiring the knowledge suggested above.

Finally, it is critical to take care of yourself. Martyrdom helps no one. Caretaking is physically exhausting and psychologically depleting. It is right to take time for yourself in ways that will refresh your body, mind, and spirit. Do not be shy about asking for help from friends, agencies, programs. Do not feel guilt about sharing the burden of dealing with a difficult vet. Know that if you don't take care of yourself you will not be able to offer the patient care typically required by vets with post-traumatic stress, brain damage, and/or all the physical, moral, and psychological wounds of war. Every family member and friend of vets would benefit from contact with other families and friends of vets. Whether it is a formal or informal group or just an individual, there is no substitute for the support available from others who have gone or are going through the same veteran re-entry process as you. It may well also be that there will be times when you would benefit from counseling from the right MHP or the right vet. This book provides the information that can allow for the selection of the "right" helper.

Returning vets need the love, support, and understanding of their family and friends. It is not an easy task to keep giving, to not judge, to understand, to have compassion and patience for someone who is very limited in their ability to give back. Knowing that your veteran's pre-service essence is buried within that damaged personality provides the fortitude to stick with them. Also, know that with help it is possible for the veteran to heal and exhibit the positive wisdom and skills that were also part of his or her military experience.

Another Vital Factor: The Community

The last chapter offered the striking contrast between how vets were treated by civilians after World War Two and after Vietnam and the dramatic differences caused by those differences. That data demonstrates how the public opinion of our wars has a huge effect on our returning warriors. This chapter goes beyond community opinions to community actions.

The activation of community resources for vets has profound consequences for the health of our vets, as well as being a boon for local communities. Americans are naturally very generous, giving people, and the opportunity to give back to vets lifts the spirits, unites the community, and provides positive services to the community. It is the opposite case with communities that are non-vet-friendly; as such they are much more likely to be plagued with problemed vets for whom expensive services are needed (i.e. hospitals, jails, detox centers, homeless shelters). Simply speaking, communities that actively support vets are enhanced rather than depleted by their veteran populations.

> *"The willingness with which our young people are likely to serve in any war, no matter how justified, shall be directly proportional to how they perceive the veterans of earlier wars were treated and appreciated by their nation."*
>
> GEORGE WASHINGTON

A leading example of positive community actions is the aforementioned Welcome Home Montrose program (www.WelcomeHomeMontrose.org). Montrose, Colorado, has mobilized its entire community to provide an ideal situation for their returning vets. The six principles their program encompasses are:

1. Open Arms Community: WHM provides a "peer network to understand challenges, provide encouragement, advice, accountability, and role modeling" for successful transitions.
2. Restore Purpose and Meaning: WHM creates "settings for meaningful employment, job training, career development, volunteering, and mentoring" to "fuel for-profit and non-profit initiatives," based on the motto: Do well by doing good.
3. Adventure, Challenge, and Recreation: WHM opens "avenues to outdoor adventures" so veterans can experience and lead adventure tourism.
4. Comfortable Environment: WHM provides physical and financial counseling to assure each vet of a comfortable place to live and work with a feasible forward-looking financial plan.
5. Health and Wellness: WHM offers local support services in therapy, counseling, and wellness, negating the

need for traveling outside their home community.

6. Restore Faith and Hope: WHM's Warrior Resource Center, occupying 4,000 square feet and formally partnered with six public entities, is where slews of volunteers join to "provide support, direction, and inspiration" in one space for vets to socialize and receive local, regional, and national resources.

While there are similar programs scattered throughout the country, the WHM program has been so successful (check out their website) that many other locations, including the Huts for Vets home community, are emulating their program as closely as possible. In Colorado's Roaring Fork Valley, the newly underway Valley Vets program is evolving through communication with the experienced WHM organizers. A "Warriors Resource Center" is planned as a one-stop resource for all the vets' needs—one place where vets can meet, mingle, and mentor, as well as learn about all the resources available to them.

Each community is unique and so needs to adapt its services to accommodate the needs of its vets. But the call is out to all communities to recognize their returning warriors' needs and meet them. Waiting for or depending on big government to pay back our country's honorable debt to our veterans has little hope and is wrong.

And it is not difficult for a community to find volunteers and finances to support vet programs. Deep sympathy and gratitude for vets abounds in the American population. People only need someone to point out the direction for them to rush to get into line. We have all heard again and again the failings of the government to meet the needs of our vets. Our citizens feel real pain over the plights of vets. But as with the failed Occupy movement, if there is no clear goal or direction, even the best intentions fail.

The organizing of the community to serve vets requires three things: education, leadership, and funding. Education refers to all the opportunities that are available for citizens to meet, mingle with, and hear from vets. In Colorado's Roaring Fork Valley, we have given plays about and performed by vets with after-play Q&As between audience and vets. We have sponsored a library program called War

Stories, in which vets and citizens meet for a number of weeks to share thoughts about relevant literature dealing with war. Any vehicle for integrating information about the needs of vets will stimulate citizens to want to meet those needs.

Community leadership only means that an individual or group needs to do the work of organizing some program for the community to support. It takes some knowledge of vets and their needs to come up with a meaningful local program. Whether it is an outdoor recreational program, a library program, a play, a seminar, or the creation of a space for multiple vet uses, a devoted, educated leadership is essential. Sometimes this has come from local governments and sometimes from community groups as unexpected as a local book club. Vets themselves can provide the primary leadership, but if not, that leadership does best by including local vets in all stages of planning and operating programs for vets. The advantages of having vets involved in any vet program is related to the already-discussed critical necessity of having vets feel the comfort and understanding they only get from other vets.

The funding of any community program for vets is typically misconstrued as a major problem. It is not. Huts for Vets has garnered substantial financial support from individuals, service organizations, local governments, businesses, and foundations, all with minimal effort. Just having our board members talk to various groups and hold programs that always offered opportunities for giving resulted in a pouring forth of contributions that testified to the desire of all those individuals and entities to help our returning vets. The public is aware of how badly our former warriors are treated by the VA. The public is aware of the magnitude of visible and invisible wounds suffered by so many vets. The public understands that the homelessness, drug addictions, anger problems, unemployability, and other such problems are rampant among our vets. Show that public that you have a way of actually helping with those problems, and you will have no trouble gaining the donations to fund good programs.

In summary, the right way to treat vets involves, first, an understanding of the sources of and individual reactions to their war experiences. Reading their literature is a major way of accomplishing this, as is interacting with them in an environment where they feel understood, appreciated, and supported. Secondly, harnessing the unique power of vets to help vets provides an efficient and effective treatment modality that should be utilized at every stage of helping vets with their transition into healthy, well-functioning civilians. And, finally, involving the local community in providing programs and services for vets is a winning strategy. This is especially true in comparison to depending on government, particularly the federal government, to provide the multifaceted support vets need. Just think of the massive failures of the VA.

Notes

You can contact Dr. Alpern with questions or comments

at www.VetsForVets.info

Chapter Four

HUTS FOR VETS
History and Procedures of a Successful Wilderness Program

Unsolicited Testimonials from the Very First 2013 Huts for Vets Trip:

"Wow! Words cannot express the impact this experience has had on my heart and soul! Wilderness therapy for combat vets is an incredible concept. What a journey! I have come further with braving my post-traumatic experiences in the last three days than I have in two years dealing with the VA. I pray that my fellow brothers and sisters in arms find their way to experience the enlightenment that Huts for Vets has provided."

"This was my first time in the wilds and it was with the best group of veterans I've ever been with. Huts for Vets and my fellow warriors saved my life over the last four

days together. Finding healing from all the things we've experienced could not have happened in a better place."

CONSTRUCTING HUTS FOR VETS

Early in 2013 I was approached by Paul Andersen, a writer and wilderness guide, but mostly a man with a mission. Paul is passionate about the beauty and enormous value of wilderness. He had long been piloting world leaders from the Aspen Institute into the wilderness for seminars. Upon learning of the plight of returning post-9/11 vets he had an idea: taking a group of vets into the wilderness for a weekend of healing.

I was approached as a mental health professional (MHP) with experience in treating veterans. Together we created a successful program that has now completed three years of multiple Huts for Vets wilderness trips with vets (www.HutsForVets.org).

This chapter offers the history and techniques of Huts for Vets, as well as the discoveries we've made along the way. The purpose is to introduce innovative, non-traditional treatment concepts that have provided a level of success very rarely matched by the currently approved treatment modalities sanctified by universities or the VA.

The Huts for Vets organization began by putting together a board comprised of veterans from every war since the Korean War who were dedicated to the well-being of vets. The original nine-person board consisted of retired officers and enlisted men and women from the Navy, Army, and Marine Corps.

The original concept was to provide a free trip into the wilderness for Vietnam and post-9/11 combat vets who had suffered physical and non-physical wounding. Beyond the exertion of the hiking and the exposure to the beauty, challenges, and teachings the wilderness had to offer, the wilderness trip, which is described below in detail, also included a collection of readings selected to accompany structured discussions.

There are a few—very few—other outdoor programs that focus on vets. Two of these are major programs of known high quality. The first is Outward Bound (www.OutwardBound.org/veteran-adventures). They run multiple-week programs, but generally only one per year. They are well-funded, having gained a 4.3-million-dollar grant from the Military Outdoor Initiative Project. That same organization also funds a number of programs through the Sierra Club (www.SierraClub.org/outings/military). They run a variety of multi-activity programs, some including hiking, archery, bonfires, and fishing. But I believe they only ran two such programs in 2015.

There are also a few animal vet-therapy programs that pair up vets with various animals, dogs, horses, and even parrots (see Charles Siebert's article "The Parrots of Serenity Park" in *The New York Times Magazine*, 1/31/2016). Many subjective reports from these will note the effect of bonding that occurs between vets during their programs, but none focus on the bond's effect of fostering healing, and none have reported formal empirical studies of their programs.

One formal study using reliable and valid measurements was conducted by Elizabeth Vella, Ph.D. That study, published in *Military Medicine,* dealt with a fly-fishing weekend for vets in Utah. When I began planning for the 2015–16 Huts for Vets study, I contacted Dr. Vella, who agreed to serve as the co-investigator for our study, which is currently utilizing the evaluating instruments and some of the techniques of her published study, with the additional scientifically enhancing aspect of the inclusion of a control group. This study is described in more detail later in this chapter.

What has been amazing has been the generosity of the community to support Huts for Vets. The Aspen Skiing Company and others have provided wonderful houses for lodging before and after wilderness trips. Restaurants have donated food for all meals. Utility companies have given us vehicles for our transportation. Sporting goods stores have provided our camping equipment. We've received donations from theaters and other local businesses whenever various needs have arisen. And then there are the wonderful citizen volunteers. Generally these volunteers have been community members who attended our events designed to educate the public on the plights and needs of our veterans. And *cash*. Large and small donations have filled our coffers and guaranteed the financial future of our programs.

The critical point here is that communities are very willing to help vets. They only need educating about the needs of veterans and a venue that makes sense to them. Americans have always been known for their generosity, and the response to HFV has proven their commitment to our vets. The takeaway message is that finances will not be a problem for communities that develop the right kind of services and programs for vets.

Rather than a chronological history of HFV's three years of experiences and our findings from a dozen such trips, this chapter will provide a description of the basic program as it has evolved, the theoretical reasons for each aspect of the program, and some suggestions as to how to adapt and modify the program to fit a variety of geographical settings. The hope is to provide guidelines to generate multiple wilderness programs across the country to serve a larger percentage of needy vets than the relatively small number now being helped by the limited wilderness programs currently available.

Program Enlisting and First Night

The vets participating in an HFV trip are enlisted through multiple means. The VA, neighborhood groups, community presentations, military installations, churches, hospitals, and word of mouth have all been used to find the vets that participate. It is important to note that vets have become suspicious of all treatment programs; we have had vets question the legitimacy of HFV ("Is this another scam of some sort?") in addition to its effectiveness. Recruitment was especially difficult before the initial successes of the first programs. Since then those vets who have attended a program have, far and away, emerged as our most effective recruitment tool. Our last group was filled almost entirely with vets who were told of the program by other vets.

In order to sign up, the potential participants complete application forms that review and document their service records; provide their medical information, including diagnoses, medications, information on their physical fitness and/or limitations, and psychological treatment history; list family, emergency, and medical contacts; and, finally, articulate their reasons for wanting to attend. All vets are eligible, although preference is offered to post-9/11 combat vets—primarily those with "invisible" post-traumatic stress, brain-injury, and moral wounds, the latter of which are inferred from a collection of psychological problems such as depression, social isolation, or devitalization. Multiple ranks and branches of service are mixed within trips. Sexes are not; males and females are taken on separate trips. Once accepted, the vets are informed of what they need to bring and prepare for, and are given a book of readings that will be the basis of the periodic "seminars" available during the trip.

The program runs from Thursdays through Sundays. On Thursday, the vets convene at donated large homes or ranches that can sleep a dozen or so. Having the participants get to know each other by living together the first evening has proved valuable in forging the bond between the vets, who mostly come from different home bases and so are strangers to each other.

That first evening begins with an informal get-together lubricated by special appetizers provided by volunteer local health-food chefs. Then, prior to dinner, the HFV staff and trip participants sit in a circle and introduce themselves. We start with the HFV board members, each of whom tells of their experiences in the service and their struggles to return to civilian life. Those authentic and moving stories precede the participating vets telling their own stories. This circle of stories has always produced feelings of belonging and appreciation that the participants are among comrades who have lived through similar experiences. Frequently we have heard that this is the first time since leaving the service that they have felt comfort, understanding, and trust from anyone…including their own families, associates, church, work, or their communities in general.

That first day ends with a communal meal and an evening of unstructured interactions. They are given details about the trip such as the lengths of the hikes, hints for packing and self-care tips (such as heavy hydration), and their freedoms and limitations during the experience. One limitation involves alcohol. Initially we served beer during the initial get-together and during the trips. The thinking had been that we wanted to treat our participants as adults who had the right to drink if they so chose. However, some vets objected to the presence of alcohol, as they were recovering alcoholics and did not want the exposure. And we experienced some behavior problems from those who drank too much and interrupted the flow of various parts of the program. These problems led to HFV deleting all alcohol from all aspects of the program.

Day Two: First Hike, First Seminars

Dawn breaks and the vets are packed up and taken to a trailhead at 9,200 feet. The trip to the hut takes roughly six hours and covers a 3,200-foot rise in elevation. Non-fit and handicapped participants are transported to the hut, avoiding that first, most demanding of hikes. It is a spectacular trail following a stream up through a beautiful forest with a multitude of dramatic views and vistas. The chance of encountering anyone else on this wilderness hike is essentially nil. There are stops and talks about the readings. We were slightly surprised, and very encouraged, by the vets' response to the readings from the first trip on (see Appendix B for a list of these readings). The vets were mostly enthusiastic about the readings, and participated by offering their responses to the ideas and themes presented. Those discussions closely resembled a seminar of graduate students, except for a higher level of life experiences and wisdom acquired over the course of their service experience. Initially it was shocking to hear the high level of discourse from the vets. That experience was one of a collection of surprises offered during the trips that led the staff to have a new level of respect for the sometimes gruff and "redneck" appearance of so many of the participating vets.

Eventually the group arrives at a field about an hour from the hut, where lunch and an hour-long seminar allow them to catch their breath. For the most part, the seminars (about three a day, each being about sixty to ninety minutes) are based on the reading contained in the notebook each participant carries. The readings are one- to three-page excerpts from works by philosophers, poets, psychologists, and various wilderness writers. During the seminars they are asked to read and then discuss various aspects of the selections. As described above, fears that the vets may not be up to such intellectual discourse have been proven groundless on each and every trip. The vets thrive on discussing ideas about wilderness, war, and the philosophical and psychological challenges of life. Vets have proven quite adept not only at grasping the proffered ideas but also at creatively dissecting and analyzing them in practical ways applicable to the lives of returning vets.

Other seminars are devoted to providing tools for trauma recovery. For example, an ex-combat Marine sergeant from the staff teaches physical-trauma-release exercise sessions, while a Korean-vet psychologist presents ways vets can "be their own therapist" through specific mindful or writing techniques.

At the hut, evenings and nights are dedicated to non-structured fun. Starting with snacks on the porch with a spectacular view and continuing through communal dinners and campfires, the vets swap stories that range from stark war horror stories to hilarious service snafus. The back-slapping, raucous laugher, and declarations of eternal brotherhood mask the underlying message that is so crucial to their recovery: that they are not alone in having witnessed and having participated in the atrocities that are germane to war; that their post-traumatic stress and other war wounds that painfully separate them from civilians are known, understood, and shared; and that they can recover and find a meaningful life after war. The compassion, empathic understanding, and camaraderie that each vet experiences is completely obvious by the end of this second day.

DAYS THREE THROUGH FOUR

Once the trip community has been well established, the rest of the trip is designed to capitalize on that amazing peer power. They are given "missions" to complete *together*, such as a difficult, compass-directed off-trail hikes following a reading about Lewis and Clark's experiences. Even their "solos" (an hour alone in a mountain spot out of sight of any others) are a peer experience, as they are followed by a group discussion of what each vet went through and thought about during that alone time.

The trip is all about learning two primary things: the magical benefit of the wilderness, and the even greater benefit of sharing time and camaraderie with fellow vets.

The lesson of the wonders and therapeutic value of the wilderness comes, first of all, from experiencing it all during the trip. The staff points it out, the readings sharpen the focus, and stories of wilderness healing abound. One staff member tells of how running in the forest was his initial salvation when suffering through his return to civilian life. Another extols the meaningfulness of her regular wilderness retreats. So the vets hear the stories and read the words of the likes of Thoreau and Abbey, but, most of all, perhaps for their first time since their discharge, have fun and release during what the Japanese call *shinrin-yoku* (forest bathing).

The lesson of the value of being with other vets who have shared their war experiences comes even more easily. It is obvious to mind and body how comfortable each vet is made by interacting with other vets. On one trip, in which a videographer was embedded to document the program, the following example of peer value came out. During a documentary interview a vet mentioned he had slept deeply for the very first time since discharge. He went on to say that being surrounded by fellow vets in the hut provided the safe comfort he needed to truly sleep. Sleep disorders are rampant among hyper-vigilant, fearful, tense vets. Though so many vets are desperately seeking a good night's sleep, there is no practical way to offer vet group sleeping as a solution. It may well be part of the reason so many troubled vets re-up (re-joining the military following discharge), seeking the general comfort they gain from being involved in a mission with people they trust, as opposed to the isolation of being with civilians from whom they feel so alienated.

Peer power is made obvious by the repeated telling of personal stories during the trip. This is a complete reversal of the classic behavior of vets becoming more and more isolated and silent over time. On the hut trips vets experience not only nonjudgmental acceptance (and, oh yes, love) for who they have become but also genuine understanding of their feelings and civilian plights. Once they realize that what they are feeling is common and not a reason for self-hate or complete pessimism about life, the stories pour forth. It is as if they are compelled to recite aloud every aspect of their moral wounds and terrors out of need of the acceptance that will render them livable.

The last morning at the hut involves the final seminar, followed by the vets filling out evaluations of each aspect of the trip, offering their opinions as to how to maximize the HFV experience. The exuberant praise and gratitude that has filled those evaluations has been dwarfed only by the verbally offered kudos: "You guys saved my life"; "I feel closer to everyone in this group than to the guys in my combat unit"; "Now I know I can make it"; "Thank you. Thank you." And then later we hear from spouses how wonderfully changed the vets are when they return home—less depressed, more motivated, more social, more communicative about their feelings, and, most of all, more hopeful about themselves and their future.

However, such accolades will not impress the huge entities that fund veterans, such as the VA, nor the universities that train MHPs. Those spontaneous statements by the vets too much resemble the "testimonials" that big pharma uses in their advertisements. It is for this reason that HFV has undertaken executing a formal study, discussed later in the chapter, to empirically validate the HFV experience. The plan is to provide the proof that will allow for a wide endorsement and duplication of the HFV experience throughout the country.

FINAL DAY: INVOLVING THE COMMUNITY

Following the much-easier hike down from the mountaintop, the vets refresh themselves at the donated home base and ready themselves for the final party and feast. That party is hosted by members of the community, usually those who have attended the HFV's War Stories program. War Stories is a regularly repeated library seminar made up of five two-hour discussions of a collection of excerpts from war literature. The participants are comprised half of community civilians and half of veterans. The goal is to educate the community through the written stories amplified by the real-life experiences told by vets and discussed during the seminars. It is these informed citizens who host and welcome the vets back from the wilderness. It is an incredibly positive ending, with the vets being honored by civilians that have acquired some insights into their experiences and lives. Each end-of-trip party can only be described as joyous.

The philosophy of HFV includes the responsibility of not only "dealing" with vets but also of educating the public about vets. The importance of this goal is obvious. Events like War Stories are in service of this goal, and HFV has used many other public forums to deliver similar messages, including sponsoring plays about vets performed by vets and participating in any local programs that serve vets or offer the community opportunities to give back to vets or understand their situation.

Involving the community in understanding, appreciating, and working for our returning veterans is essential for the entire country's health. We all need to address how we can, together, better inform the community and enlist them in paying back the vets who have sacrificed so much.

Following each and every trip, every participating vet has, through unsolicited notes, evaluations, and follow-up letters, offered the highest of praise and emotional thank-yous to the staff: "This has been a catalyst to healing and becoming whole again." "I have a new, positive view of what life can be for me." "It helped a lot to prioritize life, which can be very difficult to do with PTSD." "I am not a horrible person." But while such subjective testimonials do have some value (e.g. for future program enrollments), to gain major support from academics, government agencies, and community programs, you need objective data. Dr. Elizabeth Vella and I designed the most comprehensive vet program evaluation study to date.

The 2015 trip participants filled out five valid and reliable single-page questionnaires measuring PTSD symptoms, quality of sleep, psychological functioning, emotional functioning, and quality of life. The testing was done two weeks before, immediately after, and six weeks after the trip, thus acquiring data on their psychological health before the trip and on the effects, both immediate and long-term, the trip had on it. At the same time intervals, the same tests were administered to vets from the HFV waiting list to provide a control group. The final study data are currently being collected and analyzed. In order to obtain larger numbers of subjects and thus increase the quality of the study, HFV is planning to include additional data from the 2016 trip participants and waiting-list subjects. The study is planned to be formally presented in a leading psychology journal and will also be summarized in Paul Andersen's upcoming manual for replicating HFV-like wilderness programs. There will also be a study report available on the HFV website, www.HutsForVets.org.

The unbelievable value of these trips has been not only providing a significant healing experience for vets but also providing them with powerful tools with which they can continue to heal themselves. HFV is also developing follow-up programs for the trip participants to encourage continued mutual mentoring and allow for periodic contact with HFV for support and guidance. Future HFV plans include years of conducting and learning from these wilderness trips, as well as disseminating information about their success, along with guidance and instructions for carrying out similar programs in wilderness areas that exist throughout the country.

Opportunities for outdoor therapy programs abound. There are 59 national parks, 154 national forests, 765 wilderness areas, and thousands of state, county, and municipal parks in the U.S. Hopefully they will be used for the kind of outdoor healing for vets HFV has demonstrated is possible. It would be a step in the right direction if all city, state, and national parks and wilderness areas would provide all vets with free access. Eliminating admission fees for vets would deliver some strong messages: First, that we're paying back the vets by providing them free access to the lands they fought and sacrificed for. In addition, it could encourage vets to immerse themselves in environments proven to offer great healing potential. (For more information, check out Eva M. Selhub and Alan C. Logan's well-documented and very readable and informative 2012 book *Your Brain On Nature*.)

MORE HFV PARTICIPANT QUOTES:

"The peer group approach is exactly what I needed to realize, and admit, that I have a problem beyond my self-diagnosis, and I am ready to take it on."

"I want to again thank all the staff and participants in Huts for Vets for all their efforts, sharing, personal risks and disclosures, and feedback you gave to me in this unique group wilderness experience. I feel very overwhelmed (in a good way) with emotions and reflections from my whole life, and also very grateful to have been there in the presence of extremely cool and talented people."

"Being in this wilderness has helped me understand the value, quality and significance of these wild places. 'The further man's feet are from the earth, the less respect he has for living, growing things.' The men here have taught me that I must improve myself in order to enjoy the freedoms that we all fought for. The Huts for Vets staff is an amazing group of guys with an exceptional program that can give the combat veteran tools to be able to cope with transitions and the issues associated with PTSD, TBI, etc. This program exceeded all my expectations, and the talk, the laughter, music and general BS were truly a gift and proved that camaraderie and brotherhood cross the lines of military branch, unit, and years. Thanks for everything!"

Notes

You can contact Dr. Alpern with questions or comments at www.VetsForVets.info

Chapter Five

THREE PEER-ORIENTED TREATMENT MODALITIES
Harnessing the Power of Vets to Heal

Peer Treatment Theory

The theory behind peer-oriented therapies for vets begins with the fact that veterans struggle with talking about their feelings, experiences, behaviors, or beliefs. They believe, correctly, that "If you weren't there, you *cannot* understand." That fact, coupled with the self-alienation generated by moral wounds (having done things in combat zones antithetical to their pre-service ethics and morals), significantly hinders or simply stops them from addressing their problems and participating in most classic trauma therapies.

This leads to the fact that vets can talk about their feelings, experiences, behaviors, and beliefs *with fellow vets* who have had similar service experiences and *can* understand. Again and again, HFV and similar peer-centered programs have shown that a group of vets can form a community in which there is very high-level trust that their fellow vets understand them, do not judge them negatively for what are not uncommon combat behaviors, and share the problems of adaption to civilian existence. Furthermore, surrounding vets with their counterparts stimulates the sharing of pains, problems, and coping mechanisms. HFV sessions are replete with examples of vets opening up, frequently for the first time, and pouring out what had been bottled up since leaving the service. In fact, such sessions have sometimes needed to be cut off by switching from a simple ventilation of horror stories to a communal search for ways to deal with such memories.

This power-of-peer phenomenon is not unique to veterans. Peer-support groups have long proved to be not only *an* effective treatment but, in some cases, *the only* effective treatment; think of Alcoholics Anonymous, considered a premiere treatment of choice for alcoholics. It employs not only the peer group dynamic but also the peer mentoring aspect (i.e. sponsors), which is described below for vet groups. Consider also the fact that the now very standard cancer peer support group, which was unheard of just a decade or so ago, is now commonly available in every community. Why? Because careful research empirically proved that such groups improved the quality as well as the quantity of life. It's noteworthy that the obvious benefits of the early cancer support groups sparked the proliferation of other support groups nationwide even prior to the empirically validating research being done. The proliferation of veterans peer groups will most probably follow the "practice before proof" path taken by cancer groups.

The question then becomes: how can the power of peer group trust be translated into effective, scientifically valid therapies for veterans? This chapter attempts to answer that question with three therapy strategies currently being investigated and tested.

Peer Group Mentoring

Every sizable community in the United States is home to large numbers of vets. These vets are somewhere on the continuum of making the transition to civilian life, from very well to very badly to not at all—i.e. suicide. Add to that the fact that most veterans are highly motivated to help their struggling comrades, i.e. to "have the backs" of their fellow vets. Peer group mentoring involves formulating a group of vets in a community with the stated purpose of learning to mentor vets in the same community who are struggling with aspects of post-traumatic stress and/or brain injury.

The benefits of support groups—i.e. providing comfort, emotional nourishment, information, and like-suffering companions—have been proven to be a major therapeutic tool for many conditions. Two prime examples of this use are, as we discussed above, AA for alcoholics, drug abusers, and co-dependents, and cancer groups for each type of cancer. The international success of AA is legend; likewise, a collection of studies have demonstrated that cancer patients who participate in a support group live a higher-quality life, have better responses to therapies, require less pain medications, and live longer.

Forming such a group with vets is not difficult, as veterans love a mission. That is: for soldiers, the carrying-out of focused missions was not only a critical part of the job, but also a major mode of coping psychologically with the many incomprehensible aspects of warfare. By focusing on the specific mission at hand, soldiers were spared from dealing with the illogic of the horrors or snafus or craziness, all of which are inherent in waging warfare.

Veteran peer grouping is an exciting and dramatic activity. Vets are so relieved and comforted to be with fellow vets who speak their language and share their generally unspeakable experiences that vet groups form into dynamic communities much faster and more deeply than most support groups.

The general format and operating aspects for the group would be as follows:

The stated purpose of the group would be to teach vets how to mentor fellow vets with whatever aspect(s) of transitioning to civilian life with which they are struggling. An automatic, though unstated, purpose of the group is to provide the participants with the same kind of useful bonding and therapeutic interactions that take place in Huts for Vets or similar wilderness programs. Such programs (see Chapter Four) are known for their peer group therapeutic benefits.

The "curriculum" for the group sessions would begin with imparting the purpose of the group as guiding them to be able to function initially as "buddies." Once a trusting relationship has developed, the curriculum will teach them how to "mentor," i.e.

to help individual vets through the difficult process of successfully returning to civilian life.

The group sessions would then evolve organically, following the needs and directions of each group's participants, but would include the following activities in whatever order:

- Having participants tell their own stories to the rest of the group. This can be one or many stories. The stories could involve war stories or stories of their own re-entry process. The goal here is to provide for them the useful experience of having their stories heard by fellow vets who can uniquely understand and emphasize with the storyteller.

- Intellectually understanding the processes of changing from civilian to trained recruit, from naïve recruit to war-knowledgeable soldier, and from soldier with PTSD to functioning civilian. This education can come from selected readings, Q&As, mini lectures, or spontaneous group discussions. Differing combinations work as a function of the individual groups.

- Providing instruction on the specifics of mentoring: How to start by developing trust through simple, sharing friendship. How to encourage the mentee to tell his or her stories first to the mentor and later to loved ones. A crucial aspect of mentoring is how to handle it when the mentor is stuck about how to deal with some aspect of the mentee's functioning or behavior. Mentors need to feel confident about dealing with things from their mentee that could overwhelm them, such as suicidal indications, unbridled anger, or deep depressive behaviors. This usually involves having a connection to a knowledgeable MHP for counsel and advice. Many times, the MHP offering the mentor group serves that role for the participants. How to develop and use mental health expertise is a major part of mentor training.

- Developing a toolbox with teachable tools for dealing with various aspects of service wounds. There are many body-oriented approaches for dealing with stress that can be taught. There are writing tools, forms of journaling, that can be used for anxiety, depression, or stress reduction. There are tools as simple as lists of contacts for physical, mental, financial, educational, or occupational support specifically for veterans. The more tools the mentor can offer, the more effective mentoring can be.

The peer mentoring group should be an open rather than closed group. This means that participants can join or leave the group at any time. Closed groups are classically more intense, but the typical veteran is not prone to commit to any activity where they cannot freely choose to come or go. Also, the periodic introduction of new members allows older members to experiment with their mentoring skills.

The primary advantage of peer group mentoring is simply that it harnesses the immense power of peer understanding and support for the helping of countless numbers of veterans. It is worth underlining the proposition that PTSD is both an inevitable and honorable war wound. It will take legions of knowledgeable helpers to treat our multitudes of returning vets. Peer group mentoring allows for affordable, mass treatment for veterans by those that are most able to understand and guide their fellow warriors through the transition process.

Individual Peer Mentoring

There are going to be huge numbers of vets in need of help in locations where no such group of peer mentors, as described above, are available. And many times a vet with problems will be referred to a community-based MHP. The present individual peer mentoring therapy (IPMT) technique involves the MHP teaming up with a vet peer to treat the referred vet.

As has been repeatedly reported and documented, vets are extremely reluctant to engage in talk or, likewise, almost any other form of therapy with civilians, who they feel cannot know what they've endured, cannot identify their underlying problems, and would judge their service behaviors negatively. So even vets who are *trying* to work with and trust civilian therapists put up unconscious resistance, which can fatally undermine traditional therapy. These problems with traditional therapy are so ubiquitous that it is widely considered unethical for a civilian therapist to accept vets as patients except in cases when they fully understand and creatively modify their therapeutic approach to accommodate these problems. Here the creative accommodation involves using a non-MHP vet as a co-therapist.

In its purest form, IPMT would proceed as follows: When a vet comes to a mental health professional for therapy, the preliminary diagnostic phase would include gaining information on the vet's relevant service record. That is, his or her rank, branch of service, service occupation, combat and deployment experiences, physical and non-physical wound record, past treatments and current medications would be part of the intake information. The HFV application form available at HutsForVets.Org offers an example of pre-program information felt significant for providing optimal value for the vet. Once the MHP knows those critical facts about the vet, he or she seeks another vet with a similar background to serve as the primary agent in the therapy process.

A real-life illustration of this technique began with a call from a vet's wife who was seeking help for her very troubled husband. She did not feel the vet would ever agree to come to see any mental health professional. Her husband was a Vietnam combat vet with severe depressive and anger symptoms who was unmotivated to participate in most normal-life activities. A vet from a vet organization who was himself a Vietnam combat vet who had recovered from years of severe depression following his discharge was found. The mentoring vet then met with the MHP and through various strategies became acquainted with all the goals and techniques described above for peer group mentoring. He then approached the troubled vet with the offer of acquaintanceship/friendship, beginning with an introductory meeting in a café—*not a formal office*. Very slowly, the relationship developed, beginning with sports talks and eventually approaching conversations about the quality of life after the service. To shorten the story: The MHP never once had direct contact with the mentee vet. Rather, the MHP's role was to be available to the mentoring vet whenever the mentor felt he needed advice or direction. So when the monitoring vet reported sleep problems for his "buddy," the MHP guided him to information about sleep hygiene. When suicidal ideation was the problem, the MHP became very active in offering guidance about determining the seriousness of the threat and the specific steps to take in various scenarios. That vet, who probably would never have sought, much less been helped by, traditional therapy approaches, was helped by IPMT. He never joined any community vet activities, but after a few months of contact with his mentor, his wife happily reported living with a more relaxed, more productive, and much more positively interactive spouse.

The pattern for IPMT involves much flexibility. For instance, the role of the MHP can range from much to zero direct contact with the vet client. Also, the frequency, duration, and form of the mentor contact is completely variable depending on the situation. What is true for good traditional therapy—"create a new therapy for each client"—is even truer for IPMT.

There are legal considerations with IPMT. The question is what legal obligations are incurred by a professional mental health worker who "supervises" a non-professional (the mentoring vet) working with a "client" or "patient"? If the treated vet is considered to have entered into a professional relationship with the mental health "supervisor," then he or she has all the responsibilities due any client or patient in one's practice. This is true whether or not the supervisor is working for pay or volunteering. Furthermore, what constitutes *supervision* is defined differently in different jurisdictions. Some would consider running a peer mentoring group supervision, while others would consider it teaching. Whether it is formally done as supervision, coaching, or teaching is extremely important, as engaging in each one has significant implications for the legality of one's professional license and for determining insurance coverage. The answer is extremely complex, as the role of the MHP varies in intensity and contact depending on the needs and circumstances of the vet. Furthermore, each state and each insurance company may interpret the professional relationship differently. Therefore, before undertaking any of the peer-involved therapy activities, the MHP should check with their state-licensing entity as well as with their insurance company to determine whether their interaction with the vet is regulated by the licensing board, and how any claims would be handled by the professional's insurance. This involves a couple of phone calls or email exchanges and typically is worked out so that any of the mentoring vet therapies can be practiced with impunity. Two of the more common ways for dealing with the legal responsibility matters for the MHP involve "consulting" or "coaching," as opposed to "supervising," and the use of a legal document signed by all that specifies the role and legal responsibilities of all parties.

Many states differentiate between "supervision," "consulting," and "coaching." In general, supervision involves the supervisor having legal responsibilities for the client, as well as educational requirements of the supervisee. Thus in many states, the MHP typically will not "supervise" a lay vet doing peer monitoring. The way states define either coaching or consulting can allow for their mental health expertise to be available to the peer mentor without having legal responsibility for the client. However, it is always wise to have an attorney who is well-versed in mental health law to prepare a clear-responsibility document for all parties involved in IPMT.

MENTAL HEALTH PROFESSIONAL PREPARATION FOR SERVING VETERANS

Licensed MHPs can "legally" treat veterans regardless of their knowledge, training, or experience with veterans. But being legal doesn't translate into being capable. No one would want their dermatologist or psychiatrist to operate on their shoulders, yet to do so would be legal. Likewise, a psychologist can legally offer diagnostic and treatments services to geriatric or schizophrenic adults despite the fact that their training and experience might be with children. The point here is that if MHPs are going to offer services at any level with veterans, they have an ethical and professional need to be educated before working with that very unique population.

It has been repeatedly noted that veterans do not trust or believe that anyone—not family or pre-service friends or, especially, non-service MHP—who has not experienced their culture or actual war experiences can understand—and certainly not treat—them. So, how then can the myriads of MHPs who have not served in the military gain the understanding of vets necessary to provide them with competent services?

The answer can't be precluding the large number of trained MHPs from serving the million-plus returning vets desperately needing help. But even to provide the limited services described for the peer-based group and individual therapies, they need certain basic education.

The basic education considered necessary for any MHP working with veterans begins with a deep understanding of their idiosyncratic military culture. There are four basic sources for gaining the necessary insights into the relevant military culture:

1. Online courses teaching the culture of the services, especially the culture of combat.
2. Academic literature on the conditions most specific to returning vets (e.g. PTS and TBI).
3. Books by veterans describing their service experiences.
4. Interaction with veteran peer groups.

Appendix A is an annotated bibliography describing recommended readings in all of the above areas.

Any Internet search for "military culture" will provide a slew of online courses, many government-designed, that are free. These courses range from simply offering an understanding of the basic military rules and regulations (like rank, medals, rules of engagement) to articulating spoken and unspoken rules, morals, and ethics of combat. It behooves any MHP working with veterans to review such courses, finding those that best meet their particular educational needs.

The U.S. Department of Veteran Affairs' website *National Center for PTSD* is one major source for finding relevant and *free* online courses. Mental health professionals can google the site, log on, and click the tab labeled "Professionals." There you can look up and sign up for courses you wish to take or review. Some examples of courses available are "Understanding Military Culture," "From the War Zone to the Home Front," and "Sleep Problems with PTSD." This same site offers various questionnaires for clinical or research use.

The academic literature that MHPs need familiarity with involves, at a minimum, two areas: the diagnosis and treatment of the symptoms of post-traumatic stress, and the diagnosis and treatment of the symptoms of traumatic brain injury. MHPs vary in their experience and training with trauma and neurological matters, so it must be up to the individual to seek out the academic gaps in their knowledge base to allow them to effectively work with vets. A critical addition to the professional literature noted above is the work of Dr. Edward Tick in his two books *War and the Soul* and *Warrior's Return*. Dr. Tick has provided a whole unique approach to understanding the plights of veterans. Working primarily with Vietnam veterans, he has humanized complicated psychological concepts and suggested a collection of ways for families, communities, MHPs, and veterans themselves to, at a gut as well as an intellectual level, understand suffering vets and what they need.

The third component of a basic education for working with vets comes from reading books talented veterans have written about their experiences. Freud once opined that MHPs could learn more about the human condition and psychology through good literature than from professional writings. I certainly experienced that in my own preparation for working with veterans. I read vet books. I cried over vet books. I learned more than I can adequately describe from vet books. This book's appendix provides an annotated bibliography describing a collection of highly recommended books by vets. I can state that I learned more from reading about the experiences and emotions generated not only by warfare but also by the attempt to return to civilian life than from any other source—with the exception of the fourth cornerstone of preparing to work with vets: meeting with vet peer groups.

The experiential education I gained from peer groupings almost exclusively came from years of extended weekend wilderness trips with vets. Ostensibly I was there as the psychologist in residence, but in reality I was a shadow figure, mostly observing and taking in all the vets taught through the process of being opened up and freed by the peer experience. I did run a few hours of seminars on each trip. During the early trips, these were mostly explorative, such as asking them to review their positive and negative interactions with MHPs, or their attitudes toward drones. Only on later trips did I feel comfortable offering some psychological tools to help with common warrior symptoms. For the vast majority of MHPs, who will not go on trips with groups of peers, there are other venues available to gain veteran peer group experiences. Almost all college and university campuses now have specific vet organizations and meeting places. Volunteering with or just gaining attendance to groups of vets offers a major way to learn from vet peer groups. Similarly, many communities and non-profits provide grouping of vets. Whatever way it is accomplished, a rich education about the nature of the veteran experience is available through peer group participation.

It is certainly true that the impacts of the military, and especially of combat, cannot be fully understood if not experienced. However, this is not an excuse for MHPs working with or planning to work with veterans to not gain what education they can. Hopefully the future will provide relevant education in more of the colleges and universities for training MHPs to work with veterans. For now, the above suggestions will hopefully provide directions for currently involved MHPs.

WILDERNESS PEER THERAPIES

Back in the day, physicians would regularly treat all kinds of physical and mental ills by sending their patients to the mountains or the seaside. Then it was common knowledge that such wilderness trips were of immense benefit for conditions as wide-ranging as tuberculosis, psychasthenia (weak nerves), failure to thrive, and recovery from all manner of physical and psychological traumas. Furthermore, a myriad of writers, such as Thoreau, Abbey, Muir, and Burroughs, extolled the healing powers of nature. Google "value of wilderness" and you get 58.5 million responses. Really. More scientific testimonials for the therapeutic values of the outdoors are available through the vast amount of mind/body literature.

Today there are formal studies by major investigators scientifically documenting the positive effects of wilderness. Much of the impressive research comes from Japan. In 1982 the term *shinrin-yoku* (forest bathing: taking in, in all of our senses, the forest atmosphere) was coined to accompany a plan to involve citizens to walk in Japan's forests, which occupy 64% of their land. Since then, university and government researchers have collaborated on detailed investigations involving projects to evaluate physiological markers. For example, Chiba University collected psychological and physiological data on 500 citizen subjects, and a separate group from Kyoto has published research on another 500. These studies offer empirical evidence that *shinrin-yoku* can reduce psychological stress, depression, and hostility while increasing quality of sleep, vigor, and positive psychological states. These are also about a dozen objective studies involving twenty-four forests demonstrating lower levels of cortisol and lower blood pressure and pulse rate.

Of the many studies, I recommend review of Qing Li's "Effect of forest bathing trips on human immune function," published in *Environmental Health Preventive Medicine*, January 2015. The subjects spent two nights and three days in the forest (which is the same timeframe HFV trips use) and had blood and urine sampled twice during the trip, as well as seven and thirty days after. This was compared with control days in urban environments.

Natural killer cell (NK) activity, the numbers of NK, granulysin-, perforin-, and granzymes A/B-expressing lymphocytes in the blood, and the concentration of urinary adrenaline were measured. The results clearly showed the positive physiological as well as psychological effects of forest bathing. This important article ends with a bibliography of thirty-eight references offering similar findings from groups of scientific studies.

But beyond all this documentation are the compelling stories from vets who dramatically recount how they uniquely found peace, comfort, and healing from aspects of the wilderness. Yet most formal therapies for vets never involve the outdoors as a critical aspect. Pills, group sessions in fluorescent-lit institutional rooms, or versions of indoor talk therapy are most institutions' therapies of choice. Major mistake.

HFV takes vets up to 11,400 feet for hikes. There are other outdoor programs for vets that are based around fly fishing (141 chapters nationwide), biking, river running, horseback riding, and even trail construction and maintenance, to mention a few. The common core is obviously being outdoors with other vets on some "mission." Experience has suggested that the benefits of wilderness trips are significantly enhanced when:

- They involve a group of veterans rather than a solo experience. There is little doubt that solo time in the wilderness offers a myriad of benefits, and many vets report the value of being alone in *shinrin-yoku*. But there is clearly also benefit from the inclusion of veteran companions during wilderness outings. Peers add the possibility for that mutual support, understanding, acceptance, and mentoring so vital to healing.

- The wilderness experience includes down time, such as around a campfire, extended meals, or just BS sessions. Thus, in addition to the joint activity, be it with horses or fish, in hiking boots or swim fins, there is time for spontaneous exchanges about life. This addition mirrors the benefits of peer counseling, which is considered a prime form of PTS and/or TBI therapy. There is some follow-up to the outdoors experience. The critical bonding that comes from the shared outdoor experience needs be extended beyond the time of the outings. Plans for future or,

> better yet, regular re-groupings, continuing contact via various Internet social media, or even partial group get-togethers renders the healing more ongoing and continuous, and thus more powerful.

This chapter has offered three examples of effective peer-oriented therapy. Gaining authentic validation for these new peer-oriented therapies is extremely difficult for two major reasons.

The first is that they are inconvenient and technically difficult. MHPs, for obvious reasons, prefer treatments that can be done in their offices, preferably in fifty-minute segments. Reaching out to elicit reluctant and therapy-wary veterans for peer work is much more difficult than simply waiting for self-motivated clients with problems to set up appointments. And if you add time in the wilderness as part of the procedure, it obviously adds a whole layer of complications. The counter argument to these difficulties is that the peer treatments allow for the reaching of multitudes of needy vets, in contrast to the one-at-a-time methods employed by most MHPs. Then there is the benefit of having the vets actually performing the most effective treatments, and there are many more vets than MHPs.

The second difficulty lies in the limitations of the usual scientific methods. Erich Fromm exquisitely made this point in his book *The Pathology of Normalcy*, published posthumously in 2010. In that work he makes the point that the existing social sciences do not provide a way to gain the most important knowledge concerning mankind. These social scientists try to imitate the objective measurements of the successful natural sciences and in doing so construct studies that do not meaningfully deal with intangibles such as values or morals. Mention studying soul wounds to the typical social scientist and he or she will simply laugh at you. Yet we know soul wounds to be very real, very damaging and very prolific among vets.

Researchers dealing with human emotions and mental states, Fromm points out, choose "insignificant problems because their answers can be put into mathematical formulas rather than choosing significant problems and developing new methods for studying them." Fromm notes that the strict adherence to empirical measurable data ignores the fact that the absolute most advanced field of natural science, theoretical physics, operates not from hard data but from imaginative inferences...like Einstein wondering what it would be like to travel faster than the speed of light. So how can one "measure" the joy and relief of a wife or child crying out gratitude for the effect a wilderness program has had on their spouse or parent? We need new methods for gauging the health of things like one's soul, moral commitments, or the will to live. Our current tortuous and tangential measures do not adequately allow for the study of what forms the guts of humanity such as values, morals, or creativity.

What those two points mean is that we cannot wait for "science" to "prove" that things like wilderness and vet peers can dramatically change the health of our returning vets. We need to instead pay attention to the many subjective testimonials of the benefits of these treatments. Hardcore adherents to the scientific method may scoff at subjective testimonials, but those of us committed to alleviating the sufferings of returning vets must heed and act on these scientifically discounted testimonials.

Notes

You can contact Dr. Alpern with questions or comments

at www.VetsForVets.info

Chapter Six

PEER MENTAL HEALTH PROFESSIONALS:
How to Educate Vets as MHPs for Vets

ALL OF THE PRESENTED PEER therapies, as well as the more traditional treatment modalities, require mental health professionals (MHPs). Of course, the more traditional therapies, such as prolonged exposure, cognitive therapy, and eye movement desensitization and reprocessing, involve multiple one-on-one contacts with a trained MHP for each treatment session. And even the new peer-based therapies described in the previous chapter still require some form of difficult-to-access consulting, supervision, or coaching time from a licensed MHP.

A form of therapy that would incorporate the critical peer advantages, which are the primary focus of this book, would then ideally involve a vet who was him or herself a mental health professional.

A veteran professional who was knowledgeable about military culture and shared the basic related experiences would most be able to form the critical therapeutic alliance with veteran clients. Again, vets almost universally believe that they can be understood *only* by fellow vets who have "been there." Furthermore, veterans are highly motivated to be "on a mission," to "have the backs" of their comrades. Thus, veteran MHPs not only have the unique advantage of being accepted and trusted by their veteran comrades but also have a level of commitment to helping fellow vets well beyond that of the typical mental health professional. It is hard to overestimate the importance of the therapist's own internal motivation in the therapy process. Vets, through training, military ethics, and, especially, human bonding, are highly motivated to help other vets.

A related reason for having vets treat vets is that vets respond quite differently from civilians to anything considered "mental health." For instance, in 2015 our Huts for Vets participants were all asked to participate in a formal study of the effectiveness of the HFV wilderness trips. Though each and every vet who went on the trips was extremely complimentary about all aspects of the program, the majority of them declined to take the twenty minutes necessary to fill out the five quick psychological tests required to participate in the study. Likewise, the *Postdeployment User Guide*, a truly potentially extremely helpful book given to vets as they leave the service, is rarely read or used. That workbook, which offers amazing resources, healing exercises, and expert transition advice, is widely ignored by veterans. Both these instances show how vet culture produces a fatigue with tests and readings, which are largely associated with the "bullshit" meaningless materials (i.e. busywork) that were forced on them as GIs but which as veterans they are now allowed to happily ignore. Veteran MHPs are keenly aware of those anti-traditional attitudes and habits when it comes to mental health treatments. They are uniquely able to circumvent such attitudes and habits when diagnosing or treating their fellow vets.

Currently, the number of vets who are MHPs is much too low. Furthermore, most of those are not post-9/11 veterans, whose numbers are highest. To remedy this shortage, we need to make a massive effort to educate and train capable vets to produce the professionals needed to treat the massive numbers of our suffering returning vets. The Huts for Vets organization (www.HutsForVets.org) is developing programs to help vets become MHPs. The organization can provide scholarships funds for appropriate candidates, and is working with colleges and universities with designing and implementing such programs. Also, a few private philanthropists have joined in this educational effort. For example, private citizen and veteran Clancy Joe Herpst and his wife Linda Vitti have donated over $100,000 for such scholarships. However, these are but drops in a bucket. What is needed is public awareness of the need to educate vets to professionally help their former service colleagues heal.

TRAINING AND EDUCATION OF VETERAN MENTAL HEALTH PROFESSIONALS

The question quickly becomes: what education and training would best serve to prepare a veteran to treat other vets? Within the mental health field, there are many differing educational paths that lead to legal licenses allowing for psychological treatments. The professions involving the most extensive education and training are the doctoral programs in psychiatry and psychology. These require many undergraduate prerequisites and years of graduate training, followed by more years of internships and/or residencies. However, even such extensive training typically omits even minimal preparation for working with the very unique veteran population.

Then there is the large collection of essentially master-level trainings that permit mental health licensure. These include degrees offered by psychology, education, and social-work departments. These masters degrees all involve gaining skills in counseling, but, similar to the doctoral programs, do not offer training or education regarding the atypical culture or needs of wounded vets. Furthermore, essentially all the doctoral and masters programs include much in their curriculum that is not relevant to the treatment of veterans.

The training of physicians, nurses, and physician assistants all involve physical matters such as disease diagnosis, surgical procedures, or pediatric care not germane to wounded vets. Psychology degrees at all levels involve gaining proficiencies in areas like research design and analysis, classical theories of personality, or child development, none of which address the practical needs of veteran patients. Likewise, social-work and education degrees involve academic information and practices superfluous to the therapeutic needs of veterans. What are needed are programs in all of these disciplines directed toward practically educating their students to work with veterans.

Then there are the plethora of specialty degrees, such as drug and alcohol counseling, anger management, and life coaching, that do address some problems faced by veterans but without offering information on the specific culture or value systems of war veterans, let alone information on post-traumatic stress or brain injuries unique to post-9/11 veterans.

The point here is not to discourage veterans from seeking whatever traditional degrees suit their individual ambitions. Any degree that allows vets to work directly with other vets sans the need for a complex relationship with another professional is a good thing. However, there needs to be an appreciation of the fact that most current licensed MHP trainings omit important areas of information and skills desirable for maximum effectiveness in working with veterans. The following offers what are considered important curriculum areas *for veterans* training for licensed work with other vets.

Curriculum for Training Veteran Mental Health Professionals

The last chapter offered curriculum suggestions for MHPs who were not veterans. But even MHPs who were in the service are not exempt from the need to understand the specific culture or experiences of vets with differing fields and intensities of service. For example, a vet who was never involved in combat would profit from reading vets' books about combat. Or an Air Force vet could easily be as unknowledgeable about Marine or special force culture as most civilians. Therefore it behooves any veteran planning a mental health career focusing on vets to access his or her own areas of experience and supplement the following basic curriculum with as many of the educational suggestions for non-vet MHPs as may be appropriate for the population they are training to serve.

What follows is a listing of the information and competencies that are considered necessary to train a vet for mental health work with other vets. This list has been designed as an example of a curriculum for one or two years of a master's degree. These subjects are all ones that could be taught within psychology, social work, or education departments that offer degrees in the counseling of veterans. It would also be extremely useful if the courses could be available through local community colleges as well as through online courses. In this way, family members or vets themselves could educate themselves on the topics of most relevance to them.

Veteran Mental Health Suggested Proficiency Areas:

1. Diagnosis: Establishing familiarity with the "American Psychiatric Association's, Diagnostic and Statistical Manual of Mental Disorders", which reports on all mental health conditions with their courses and traditional treatments (including psychopharmacological). The emphasis should be on conditions common to veterans, such as post-traumatic stress and traumatic brain injuries.
2. Treatment: Studying classic treatments, such as cognitive or Pennebaker therapies as standardized on veteran populations, as well as the newer therapies as described in this book. The emphasis should be on treatments for trauma, again focusing on those most appropriate for veterans.

3. Literature Competency: Learning the concepts necessary to allow for critical evaluation of the research literature regarding the treatment of veterans. This does not mean acquiring the skills necessary to design, produce, or analyze studies, but rather the abilities to evaluate the usefulness of reported studies.
4. Suicide: Learning to determine suicide risk and strategies for dealing with it. Suicide and other serious conditions suffered by vets can be assessed with various psychological tests standardized on vets. Skills with psychological testing focused on veteran issues need be part of the veteran MHP's education.
5. Therapy Interactional Skills: Gaining competency in individual and group therapy interpersonal skills, such as rapport building and the establishment of therapeutic alliances consistent with the ethical practices of counseling, including balancing the therapeutic relationship with the social relationship.

6. Referral Competencies: Understanding two critical areas for referral: the knowledge of when a situation justifies or mandates referral to a specialist in a physical or psychological health area, and also of referrals to agencies or programs offering services to veterans.

Like all subjects, the above subjects can be taught at many levels, actually from secondary school through post-graduate study. The correct level for any given course must be determined by the agency offering the education. The intensity and comprehensiveness of the course will certainly differ depending on whether it's designed for family members wanting to learn about their wounded warrior or for those preparing for a career as a mental health professional.

The above six areas are those considered basic for veterans working toward careers in the care of veterans. The list of subjects required for non-veterans offered in Chapter Five provides an additional collection of subjects to be considered by those seeking understanding of returning veterans.

Once again, we should underscore the point that it is local communities that are best positioned to and most capable of helping veterans. In this case it would involve gaining the cooperation of local educational agencies, such as community colleges, to offer online or in other easily accessible formats the information covered by both the veteran and non-veteran subject lists. It may be difficult to have institutions of higher education develop and offer actual masters programs for vets, but the offering of courses in the aforementioned subjects should be easily implemented.

In summary, it would be extremely advantageous to both suffering veterans and motivated-to-help veterans to have more MHPs who were themselves veterans. This chapter described a practical curriculum for training such professionals. In addition, many of those subjects necessary for the training of competent MHPs are also of major interest to others—namely, citizens working with vets in any capacity, members of veteran families, and vets themselves. Because of this wide interest, it is suggested that the subjects listed also be made available to those groups at levels consistent with their interest in and interactions with vets. It is all very doable. It requires only the activation of the underlying desire to help vets, which resides in many, many citizens and community entities.

Notes

You can contact Dr. Alpern with questions or comments at www.VetsForVets.info

Chapter Seven

SUGGESTIONS AND CONCLUSIONS
Specific Help Suggestions for Vets, Friends and Family, and MHPs

AIMED AT VETERANS, THEIR FAMILY and friends, and the mental health professionals (MHPs) who treat them, this book has been crafted to serve as a manual for effectively serving our returning veterans. The following suggestions for each group of targeted readers have been distilled as a practical summarization of the information, findings, experiences, and research reviewed in the previous six chapters.

FOR VETERANS

1. Do not lose hope. You can become a psychological healthy asset to your loved ones, your community, and yourself.

Many vets believe they cannot be understood, helped, or forgiven—least of all by themselves. But there is overwhelming evidence that there are, through contact with certain professionals, comrades, programs, and communities, ways to successfully transform desperate, unhappy, or poorly functioning vets into healthy citizens. You need only recognize this possibility and seek out the successful people and methods to begin your own transformation into the person you would aspire to be.

2. Discover the benefits of the outdoors, the wilderness, and your physical self to deal with the negative residues of service.

Our great outdoors has repeatedly proven an extremely powerful source of therapy for legions of troubled vets. Many vets have discovered running, hiking, skiing, biking, photographing, and the like to be their primary and most effective therapy. Some engage in these solo, while others prefer having that experience with others: fellow vets, friends, or like-minded outdoor enthusiasts. You can contact organizations like Outward Bound, Project Healing Waters, Higher Grounds, the Sierra Club's Military Outdoors, the Army's Warrior Adventure Quest, or Huts for Vets. It is typically easiest and most useful to contact nature or outdoor sporting groups in your own geographical area. Essentially every community offers biking, fishing, and hiking clubs, and similar organizations. Seek out the one that has the most appeal and give it a try. If the first try doesn't work for you, be willing to investigate another group. Whatever medium gets you moving outdoors will provide many more benefits than you can imagine. Hard though it may be to believe, our outdoors can provide healing far beyond that offered by many expensive—and hard-to-access—traditional therapies. So step outside.

3. Explore time with other veterans, especially with ones who have experiences closest to your own service.

Spending time with similarly experienced vets frees you from feeling misunderstood, unappreciated, and unforgiven, and offers a host of other feelings many vets feel are not available from friends, family, their community, or even professional therapists. Typically, the ability to share time—to say nothing of sharing feelings, fears, opinions, service stories, and symptoms—has absolutely been proven over time to be the single most freeing and healing form of therapy for legions of vets. Spaces for vets to meet and congregate are available in most every community. Explore programs, institutions of learning, churches, buddy programs, clubs, and the VA to find the right vet or group of vets for you.

Family and Friends

1. First and foremost, become as familiar as possible with the culture and experiences of your veteran loved one. Your ability to know, as well as be trusted by, vets comes with learning the culture of the military and the kinds of experiences vets have endured.

It is commonplace for family and friends to not be able to gain the above information directly from a cared-for vet. Most veterans, especially combat vets, simply do not trust that what they have undergone, and the things they have done, can be understood or accepted by those who have not lived it. But the more the vet sees that you are working hard to understand and accept them, the more likely they are to open up to you, accept help from you and others, and simply heal. Chapters Five and Six both offer specific methods for acquiring the information sought. Do the work. It is not easy—for instance, reading veterans' books about war experiences can be painful, very painful, but not as painful as living the experiences described.

2. Become active in your community to provide services for vets and educate the community about the sufferings of vets, as well as about their avenues for healing.

There is no end of ways of involving your community to better understand and serve vets. This partial list demonstrates only the depth and variety of ways you can encourage your community to be more and more "vet friendly":

- Establishing library groups that read and discuss veteran literature (best with some vet participants)
- Establishing meeting places or activities for groups of vets
- Encouraging local institutions of learning to offer courses and degrees for and about vets

- Getting businesses to contribute money and services tailored for vets

- Calling on businesses to offer jobs, perhaps with special trainings or accommodations, to wounded vets

- Becoming politically active to encourage courts to maintain special procedures for dealing with vets who violate laws

- Appealing to churches and other religious groups to welcome vets with free services designed to meet whatever special needs are not met by other local agencies

- Joining and supporting a local group dedicated to whatever veteran services you feel able to contribute to

3. Learn how to be patient. Impatience is painful for those who have it and problematic for those who are exposed to it. Patience is extremely useful and need not equate to "long-suffering."

Patience is most easily expanded when progress is seen. You can learn to use the hope of even small steps to fortify your tolerance for dealing with very difficult personality problems. That is why it is so important for caretakers of vets to be knowledgeable about what treatments work. Such knowledge allows them to have hope and guide their vets toward the right programs, which cultivates patience. Seeing light at the end of the tunnel promotes patience, whereas hopelessness and helplessness promotes impatience.

4. Take care of yourself. Know that taking care of yourself is essential. It is crucial to find ways to recover from ongoing caretaking. The need to take care of yourself may not be understood by others who don't understand or appreciate what you are coping with. Many simply do not recognize the difficulties of taking care of those without obvious physical handicaps, when in fact, living with and taking care of the psychologically wounded is usually much more difficult and complicated than dealing with physical wounds.

Taking care of yourself might require vacations, obtaining physical or psychological help, or perhaps even ending the relationship. One important way to take care of yourself is by getting to know, and sharing your situation with, other veteran caretakers. Other vet caretakers do understand what you are dealing with and are able to offer invaluable support and information. Local veteran programs can be a source for getting you in touch with nearby families. Online social sites can be searched for contacts that suit your particular needs. Know that underestimating your need for taking care of yourself will almost certainly render you a liability, rather than a help, to your cared-for vet.

MENTAL HEALTH PROFESSIONALS

1. Be aware of your strengths and weakness in offering services to any veteran or vet programs. Continually upgrade your knowledge of and practical experiences with vets.

Realize that veterans come from a very unique military culture and can suffer from service traumas very different from the civilian traumas you may have experience with. Assess your own specific knowledge of and experience with vets. Do not provide services you are not qualified to offer. Commit to increasing your veteran knowledge base with the specific educational techniques described in Chapters Five and

Six (e.g. reading vet literature, spending time with vets in various group programs, taking continuing education courses on military culture).

2. Carefully evaluate any classic diagnostic or treatment methods you employ with vets to make sure they have been rigorously proven to be reliable and valid for the specific veteran population. That does not mean that you should not consider new treatment modalities that may not yet have been tested by rigorous scientific methodology. If, for instance, you wish to implement the new peer-oriented treatment methods described in Chapter Five, you can contribute your valuable experiences and data to the literature regarding their usefulness.

Many of the treatments and techniques empirically validated for civilians seem, but are not, applicable to veterans. Professional ethics require that you only employ systems appropriate to the population served. There are both classical and new psychological techniques whose value for veterans have been demonstrated. Review appropriate methods and then gain skills in their application, especially with those involving veteran peers or veteran peer groups. You can also make significant contributions to the vet knowledge base by conducting and reporting research on diagnostic and treatment modalities for veterans.

3. Extend services and expertise to veterans and their families in a variety of venues.

Knowledgeable MHPs are sorely needed as individual and group therapists for vets *and their family members*. Volunteering as a consultant, coach, or supervisor for vet programs offers a way to pay back some of our debt to those who have sacrificed so much for our country. Opportunities abound for MHPs to consult with businesses, courts, learning institutions, law enforcement agencies, and community programs about vets. For those who have acquired the necessary skills, teaching other professionals through formal courses or supervision is important. I can testify that work with and for veterans has absolutely been some of the most personally rewarding of a half century of professional work.

CONCLUSION

Our servicemen have suffered more casualties from suicide—about twenty-two per day, according to the study released by the VA—than from combat in all our post-9/11 wars put together. That single statistic screams the conclusion that our society is failing our wounded veterans. How we have been treating our veterans is a national disgrace. And it need not be that way.

Opportunities and information are available that would drastically reduce the entire range of painful symptoms and maladaptions of our returning soldiers, and lead those same veterans into new lives as prideful, positive contributors to their families, their communities, and themselves.

Notes

You can contact Dr. Alpern with questions or comments at www.VetsForVets.info

Acknowledgments

WITHOUT ANY DOUBT, MY PRIMARY thanks go to the many, many vets who have taught me, frequently at great emotional expense, about their lives. They have told me of their service experiences—both those that generated personal pride and those riddled with deep shame and embarrassment. They honestly shared their feelings, beliefs, and fears, which is something completely antithetical to the natural emotional wounds they've suffered. I heard very little bragging or heroic self description. Rather, for the most part I watched them courageously exposing their deepest secrets, not particularly to educate me but instead to generously give their fellow vets personal authenticity in order to allow their comrades to understand and re-evaluate their own demons. I would award these heroes with a medal that does not yet exist: a medal of psychological valor.

Next I acknowledge all those who guided me through the process of producing this book. Elizabeth Cameron, my super-duper yet compassionate editor, went way beyond substantive editing. She would, for instance, do her own research to update some of the book's statistics. But mostly it has been her gentle but very firm and extremely knowledgeable writing, organizational, and readership skills that served to upgrade the quality of the project. I always felt she was as passionately committed to distributing the messages of the book as I. Thank you, Elizabeth.

There have been many others, friends and family, who read for, advised, and encouraged me. The Price Foundation listened to my ideas and read early drafts and then awarded me a grant that allowed me to not only write the book but also produce, design, distribute, and market it. Their help will never be forgotten. My wife, Carol, and my editor friend Steve Alldredge of "Steve Stories" provided needed critiques and suggestions throughout the process. I am deeply indebted to them for their help.

Finally, a very special acknowledgement must go to my immigrant parents, who passed long before this book was conceived. But, like many of their culture, they so valued education that their lives were dedicated to making whatever sacrifices were necessary to educate their children far beyond their own limited schooling. They would proudly save every professional writing of mine like jewels, even though they were essentially beyond their educational abilities to understand. But this book, this book I know they would read and understand and so approve of that they would be *kvelling*.

Appendix A

Annotated Bibliography: Recommended Veteran Writings

POWDER:
Lisa Bowden & Shannon Cain, 2008

By 2006, over two thousand women who fought in Iraq or Afghanistan were awarded Bronze Stars and 1300 COMBAT Action Badges. Yet they are mostly unheard and unheralded. This book offers the writings of nineteen women veterans who beautifully describe soldiers' society, military history, and combat through the eyes of women who lived it from Vietnam to Iraq.

SOUL REPAIR:
Rita Nakashima Brock & Gabriella Lettini, 2012

Moral injuries are the most devastating of the psychological wounds. War promotes behaviors that lead to deep self-hatred. The four stories in this book clearly teach how self-loathing, disgust, unworthiness, regret, and grief play out, affecting the vet and all their relationships in ways not readily recognized as based in self-alienation. After the stories the book offers powerful soul-repairing tools.

THANK YOU FOR YOUR SERVICE:
David Finkel, 2013

Finkel's earlier book *The Good Soldiers* was an account of being embedded with the 2-16 Infantry Battalion during the Iraq surge. In this book he follows some of the same men after they returned home. This extremely talented writer's objective vantage point provides a description of the recovery after war more vividly and understandably than most vets' subjective views would allow.

THE WARRIORS:
J. Glenn Gray, 1959

Right after receiving his doctorate in philosophy in 1941 Dr. Gray became a private in the U.S. army. In 1959, some fourteen years after discharge, he began this analysis of what he learned about how soldiers evolve during brutal battle and how they rationalize their grossly altered ways of believing and acting. This is the book classically given college students to learn about the effects of war.

STANDING DOWN:
The Great Books Foundation, 2013

This collection of forty-four fiction and nonfiction writings covers aspects of warriorhood from *The Iliad* to the post-9/11 era. It was assembled to provide discussion topics based on each reading and includes questions for discussion. A fine collection of very readable works that provides a wide-ranging education on many aspects of the military and post-military experience.

ONCE A WARRIOR—ALWAYS A WARRIOR:
Charles W. Hoge, M.D., 2010

Subtitled *Navigating the Transition from Combat to Home—Including Combat Stress, PTSD, and mTBI*, the book does exactly that. Based on good research and written in user-friendly language, it provides a handbook for veterans as well as a text for mental health workers. A psychiatrist who was in the fray and continued his dedication to veterans after returning home, he represents the kind of mental health professional that veterans can tell has walked the walk and thus is capable of talking the talk.

REDEPLOYMENT:
Phil Klay, 2014

Klay, a retired Marine who served in Iraq, is also an educated, renowned writer. This book is truly a classic. Lauded by critics, this book spent time in the high reaches of the *New York Times* Best Seller list for good reason. The twelve fictional stories that make up the book are accompanied by such fine writing that you're unaware you're being educated about war at levels not available in most war books.

DON'T I HAVE THE RIGHT TO BE ANGRY?:
Howard J. Lipke, Ph.D., 2013

> Anger is perhaps the most dangerous and dysfunction-causing of the PTSD symptoms. This book describes anger in its many forms (e.g. depression being anger turned inward). But most usefully, this small book offers a set of tools developed by Dr. Lipke from his extensive practice dealing with veterans' anger attacks. His HEArt program is a key element in various inpatient PTSD units, but the insights and techniques are equally applicable by veterans themselves or by family members wanting ways to help.

WHAT IT IS LIKE TO GO TO WAR:
Karl Marlantes, 2011

> A Vietnam Marine lieutenant, Rhodes Scholar, and recipient of the Navy Cross, the Bronze Star, two Commendation Medals for valor, a couple of Purple Hearts, and ten Air Medals eloquently writes about what the combat veteran thinks and feels. This was seven times listed as "best book of the year"

for the very good reason that it give profound insight into the psyche of combat veteran.

NO REST ELSEWHERE:
Allen G. Orcutt, 2010

Mostly poems, some letters, some journal notes, with images all beautifully documenting the author's Vietnam experiences from 1968, as a combatant, through 2008, as a returning visitor sponsored by the charitable Dove Foundation (www.DoveFund.org). This slim volume is difficult to find but moving and insightful.

FOR LOVE OF COUNTRY:
Howard Schultz & Rajiv Chandrasekaran, 2014

This book presents two sets of heroic stories. The first set recounts combat heroics with wonderfully moving details. The second set describes how returning veterans have replicated those heroics back at home. The evolution of veterans into community leaders and community heroes answers the question of how skills learned and honed in battle can so admirably serve in peace.

TEARS OF A WARRIOR:
Janet J. Sehorn & E. Anthony Sehorn, 2008

He was an Army officer in Vietnam who returned with both physical and, even more challenging, psychological combat scars. She was his psychologist wife who, for thirty years following his service, lived, coped, learned, and studied the effects of and treatments for PTSD. Together they have written, in a very accessible style, an amalgamation of solid resesarch and personal stories that provide readers with very practical advice on what can be expected and what can be done to deal with and recover from post-traumatic stress.

THE THINGS THEY CANNOT SAY:
Kevin Sites, 2013

Eleven servicemen say the things they cannot say to journalist Kevin Sites. The gripping war stories are told though journals, emails, interviews, and poems, each ending with an updated postscript offering true follow-ups to the true stories. Storytelling has redemptive power for the storyteller, but in this outstandingly authentic form it has tremendous power to offer civilians dramatic insights into the souls of wounded veterans.

WAR AND THE SOUL:
Edward Tick, Ph.D., 2005

Dr. Tick is the preeminant veteran psychologist. This is no ordinary brilliant book. Bravely employing the "unscientific" word *soul*, Tick reviews history, mythology, psychology, and on-site observations to deeply understand PTSD beyond anything offered by medicine, philosophy, psychology, and religion. The amalgam he so clearly presents uses all these disciplines to provide gut-level understanding of veterans and ways toward healing. This is truly a "must read" for anyone caring about vets, especially Vietnam vets.

WARRIOR'S RETURN:
Edward Tick, Ph.D., 2014

Incorporating all the transcendent insights from his solidly researched book *War and the Soul* regarding the warrior's path, this book, simply stated, offers the cures for PTSD. Dr. Tick refutes the concept of it being a psychiatric condition and rather considers it a natural consequence of war experiences. He then goes on to review multicultural healing rituals leading to a collection of powerful techniques for healing veterans and reintegrating them into civilian communities. He believes and demonstrates how vets should and can have an honored place in society displaying their battle-won characteristics of strength, dedication, resilience, and hard-earned wisdom.

Appendix B

Huts for Vets Reading List

"Nature Prayer," Edward Abbey

"Over the Bitterroots, September 1–October 6, 1805," from Chapter 23, Steven Ambrose (describing Lewis and Clark's experiences)

"Veterans Expeditions to Wilderness and Regaining Health," Stacy Bare

Field Guide to Wilderness Survival, excerpt, Tom Brown

Man's Search for Meaning, excerpt, Viktor Frankl

"Walking," quote, Soren Kierkegaard

Gift From the Sea, excerpt, Anne Morrow Lindbergh

The Singing Wilderness, excerpt, Sigurd F. Olson

Walking It Off: A Veteran's Chronicle of War and Wilderness, excerpt, Doug Peacock

"Speaking of Wellness: Green Exercise: A Walk in the Woods," John Swartzberg, M.D.

"The Tragic Blessing of Combat Vets," Auden Schendler

"St. Crispin's Day Speech," excerpt *Henry V*, William Shakespeare

"Indian Life," Luther Standing Bear

"I Went to the Woods," from *Walden*, Henry David Thoreau

"Solitude," from *Walden*, Henry David Thoreau

On the aftermath of World War Two, Laurens van der Post

"Brotherhood," Sergeant Dena Price van den Bosch

"Lines Written in Early Spring," William Wordsworth

"As Is," Jonathan Young

Bibliography

Armstrong, Taylor. *Hiding From Reality: My Story of Love, Loss, and Finding the Courage Within.* New York, NY, Simon & Schuster, 2012.

Bettelheim, Bruno. *Individual and Mass Behavior in Extreme Situations.* Journal of Abnormal and Social Psychology, 38:417–452

Biank, Taya. *Undaunted,* New York, N.Y. Penguin Group, 2014

Blum, L. *Treating wounds you can't see.* (Internet). 2008: http://www.washingtonpost.com/wp-dyn/content/story/2008/06/ST2008063000520.html

Bowden, Lisa & Cain, Shannon. *Powder: Writing by Women in the Ranks, from Vietnam to Iraq.* Kore Press, Inc., Tucson AZ, 2008

Brooks, David. *The Moral Injury.* New York Times, 2/17/15

Brock, Rita and Lettini, Gabriella. *Soul Repair.* Boston, MA, Beacon Press, 2014

Finkel, David. *Thank You For Your Service.* New York, NY, Sara Crichton Books, 2013

Fromm, Erich. *The Pathology of Normalcy.* New York, NY, American Mental Health Foundation, 2010

Gray, J. Glenn. *The Warriors.* New York, N.Y. Harcourt Brace, 1959

Hoge, Charles W. *Once a Warrior—Always a Warrior.* Guilford, CT, Globe Pequot Press 2010

Holmstedt, Kirsten, *The Girls Come Marching Home*, Mechanicsberg, PA, Stackpole Books, 2011

Holmstedt, Kirsten, *Band of Sisters*, Mechanicsberg, PA, Stackpole Books, 2007

Iwasakiy. Y. *Counteracting stress through leisure coping: A prospective health study. Psychology, Health & Medicine.*2006 11(2): 209-220

Kearney, David J. & Simpson, Tracy L. *Broadening the Approach to Posttraumatic Stress Disorder and the Consequences of Trauma.* The Journal of the American Medical Association. 2015;314(5):453-455

Keatley, Mary Ann & Whittemore, Laura. *Recovering from Mild Traumatic Brain Injury.* Boulder Colorado, Brain Injury Hope Foundation. 2009

Klay, Phil. *Redeployment.* New York, NY, The Penguin Press, 2014

Li, Qing. *Effect of forest bathing trips on human immune function.* Environmental Health Preventive Medicine, 2015, 2010 Jan; 15(1): 9–17.

Lipke, Howard, *Don't I Have the Right to Be Angry?* Good Looking Software, Inc. Wheeling, IL, 2013

Marlantes, Karl, *What It Is Like To Go To War.* Grove Press, New York, N. Y. 2011

Mayo Clinic. "Post traumatic stress disorder: symptoms."

Ohlrich, Warren. *10th Mountain Hut Guide.* Woody Creek, CO, Peoples Press, 2011

Orcutt, Allen. *No Rest Elsewhere.* Toledo, OH, The Dove Fund, 2010

Pennebaker, James. *Writing to Heal.* Oakland, CA, New Harbinger Publications, 2004

Sehorn, Jane & Sehorn, E. Anthony. *Tears of a Warrior.* Team Pursuits, Ft. Collins, CO, 2008

Schultz, Howard & Chandrasekaran, Rajiv. *For Love of Country: What Our Veterans can TEach us about Citizenship, Heroism, and Sacrifice.* Vintage Books, New York, NY, 2015

Selhub, Eva M. & Logan, Alan C. *Your Brain on Nature: The Science of Nature's Influence on Your Health, Happines and Vitality.* New York, NY, 2013

Seligman, ME. *Learned Optimism: How to change your mind and your life.* First Vintage Books edition. New York, NY: Random House; 2006

Siebert, Charles. *The Parrots of Serenity Park.* The New York Times Magazine, 1/31/2016

Sites, Kevin. *The Things They Cannot Say.* New York, N.Y. Harper/Collins, 2013

Stander,Valerie, Kraft, Heldi, Xiong, Linda, Larson, Gerald. *Postdeployment User Guide: Transition Workbook for Combat Veterans.*

Navel Health Research Center, San Diego, CA. undated.

Standing Down, Chicago Ill. The Great Books Foundation. 2013

Tedeschi RG, Calhoun LG. *The Post-traumatic Growth Inventory: Measuring the positive legacy of trauma.* Journal of Traumatic Stress. 1996; 9 (3) :455-471

Tick, Edward. *War and the Soul.* Weaton, IL, Quest Books, 2005

Tick, Edward. *Warrior's Return.* Boulder CO, Sounds True, 2014

VA Bonuses Paid Amid Scandals. USA Today, 11/11/15

Van Der Kolk, Bessel. *The Body Keeps the Score.* New York, NY, Viking, 2014

Vella, E.J., Milligan, B., & Bennett, J.L. (2013). *Participation in outdoor recreation program predicts improved psychosocial well-being among veterans with post-traumatic stress disorder: A pilot study.* Military Medicine, 178(3), 254-260

Wilson, Timothy. *Redirect.* New York, N.Y., Little, Brown and Company, 2011

Index

A

B

C

D

E

F

G

H

I

N

O

U

V

W

About the Author

Dr. Gerald Alpern began treating veterans at Valley Forge in 1954 while serving in the military as a neuropsychiatric technician. His current passion for helping vets has been fueled by his post-service knowledge that his military job of administering insulin shock therapy was both ""inhuman and ineffectual." This conclusion had been published years earlier in the most highly respected medical journal, *The Lancet.*

Following his Korean War service, Dr. Alpern began his professional career at Indiana University Medical School, where he spent fifteen years doing clinical work, teaching, and research. He left his tenured full professorship, moved to Colorado in 1975, and began his private clinical psychology practice and continuing research.

Dr. Alpern's practice over the next thirty-five years included the treatment of a number of psychologically wounded vets. Until 2012 his results with vets, similar to that of his colleagues, were, at best, only minimally successful and, at worst, simply failures. Then things changed. This book tells the story of that change.

This book is the result of Dr. Alpern's extensive post-2013 experiences and research with veterans. It explains why so many treatments offered to vets fail and why they continue to be used, especially by various large government agencies and university clinics. The book then offers innovative treatments that have been tested and found to be extremely effective, cost little, and can be implemented within local communities or even by veteran's families without federal government or sanctioning universities' support.

CPSIA information can be obtained
at www.ICGtesting.com
Printed in the USA
LVOW04*0813100516
487512LV00003B/3/P